Springer Transactions in Civil and Environmental Engineering

Editor-in-Chief

T. G. Sitharam, Indian Institute of Technology Guwahati, Guwahati, Assam, India

Springer Transactions in Civil and Environmental Engineering (STICEE) publishes the latest developments in Civil and Environmental Engineering. The intent is to cover all the main branches of Civil and Environmental Engineering, both theoretical and applied, including, but not limited to: Structural Mechanics, Steel Structures, Concrete Structures, Reinforced Cement Concrete, Civil Engineering Materials, Soil Mechanics, Ground Improvement, Geotechnical Engineering, Foundation Engineering, Earthquake Engineering, Structural Health and Monitoring, Water Resources Engineering, Engineering Hydrology, Solid Waste Engineering, Environmental Engineering, Wastewater Management, Transportation Engineering, Sustainable Civil Infrastructure, Fluid Mechanics, Pavement Engineering, Soil Dynamics, Rock Mechanics, Timber Engineering, Hazardous Waste Disposal Instrumentation and Monitoring, Construction Management, Civil Engineering Construction, Surveying and GIS Strength of Materials (Mechanics of Materials), Environmental Geotechnics, Concrete Engineering, Timber Structures.

Within the scopes of the series are monographs, professional books, graduate and undergraduate textbooks, edited volumes and handbooks devoted to the above subject areas.

Hemanta Hazarika · Stuart Kenneth Haigh ·
Haruichi Kanaya · Babloo Chaudhary ·
Yoshifumi Kochi · Masanori Murai ·
Sugeng Wahyudi · Takashi Fujishiro
Editors

Sustainable Geo-Technologies for Climate Change Adaptation

Editors
Hemanta Hazarika
Kyushu University
Fukuoka, Japan

Haruichi Kanaya
Kyushu University
Fukuoka, Japan

Yoshifumi Kochi
K's Lab Inc.
Yamaguchi, Japan

Sugeng Wahyudi
NITTOC Construction Co., Ltd.
Tokyo, Japan

Stuart Kenneth Haigh
University of Cambridge
Cambridge, UK

Babloo Chaudhary
National Institute of Technology
Karnataka Surathkal
Mangalore, India

Masanori Murai
Shimizu Corporation
Tokyo, Japan

Takashi Fujishiro
Geo-Disaster Prevention Institute
Kitakyushu, Japan

ISSN 2363-7633   ISSN 2363-7641 (electronic)
Springer Transactions in Civil and Environmental Engineering
ISBN 978-981-19-4073-6   ISBN 978-981-19-4074-3 (eBook)
https://doi.org/10.1007/978-981-19-4074-3

© The Editor(s) (if applicable) and The Author(s), under exclusive license to Springer Nature Singapore Pte Ltd. 2023
This work is subject to copyright. All rights are solely and exclusively licensed by the Publisher, whether the whole or part of the material is concerned, specifically the rights of translation, reprinting, reuse of illustrations, recitation, broadcasting, reproduction on microfilms or in any other physical way, and transmission or information storage and retrieval, electronic adaptation, computer software, or by similar or dissimilar methodology now known or hereafter developed.
The use of general descriptive names, registered names, trademarks, service marks, etc. in this publication does not imply, even in the absence of a specific statement, that such names are exempt from the relevant protective laws and regulations and therefore free for general use.
The publisher, the authors, and the editors are safe to assume that the advice and information in this book are believed to be true and accurate at the date of publication. Neither the publisher nor the authors or the editors give a warranty, expressed or implied, with respect to the material contained herein or for any errors or omissions that may have been made. The publisher remains neutral with regard to jurisdictional claims in published maps and institutional affiliations.

This Springer imprint is published by the registered company Springer Nature Singapore Pte Ltd.
The registered company address is: 152 Beach Road, #21-01/04 Gateway East, Singapore 189721, Singapore

Organization

Executive Committee

Chairman

Prof. Hemanta Hazarika, Kyushu University, Japan

Co-Chairmen

Prof. Gopal Santana Phani Madabhushi, University of Cambridge, UK
Prof. Kazuya Yasuhara, Prof. Emeritus, Ibaraki University, Japan
Prof. Dennes T. Bergado, Prof. Emeritus, Asian Institute of Technology, Thailand

Secretaries

Dr. Yoshifumi Kochi, K's Lab Inc., Japan
Dr. Babloo Chaudhary, National Institute of Technology Karnataka Surathkal, India
Mr. Takashi Fujishiro, Geo-Disaster Prevention Institute, Japan

Co-Secretaries

Dr. Masanori Murai, Shimizu Corporation, Japan
Mr. Tsuyoshi Tanaka, Tokyo City University, Japan

Members

Mr. Tadashi Akagawa, Japan Foundation Engineering Co., Ltd., Japan
Prof. Yasuhide Fukumoto, Kyushu University, Japan
Mr. Hideo Furuichi, Giken Corporation, Japan
Mr. Shinichiro Ishibashi, Nihon Chiken Co., Ltd., Japan
Dr. Tomohiro Ishizawa, National Research Institute for Earth Science and Disaster Resilience, Japan
Dr. Yusaku Isobe, IMAGEi Consultant Corporation, Japan
Prof. Haruichi Kanaya, Kyushu University, Japan
Mr. Hiraoki Kaneko, Japan Foundation Engineering Co., Ltd., Japan
Mr. Daisuke Matsumoto, Japan Foundation Engineering Co., Ltd., Japan
Mr. Yoshikazu Ochi, Kawasaki Geological Engineering Co., Ltd., Japan
Mr. Seiichiro Oiyama, Giken Corporation, Japan
Mr. Taisuke Sasaki, Nihon Chiken Co., Ltd., Japan
Dr. Tadaomi Setoguchi, Former Vice Mayor, Sasebo City, Japan
Prof. Yoshihisa Sugimura, Kyushu University, Japan
Dr. Sugeng Wahyudi, NITTOC Construction Co., Ltd., Japan
Dr. Naoto Watanabe, KFC Ltd., Japan
Mr. Shigeo Yamamoto, Chuo Kaihatsu Corporation, Japan

Treasurers

Ms. Mika Murayama, Kyushu University, Japan
Mr. Takashi Fujishiro, Geo-Disaster Prevention Institute, Japan

Technical Committee

Chairman

Prof. Haruichi Kanaya, Kyushu University, Japan

Co-Chairman

Prof. Yasuhide Fukumoto, Kyushu University, Japan

Secretary

Dr. Babloo Chaudhary, National Institute of Technology Karnataka Surathkal, India

Co-Secretaries

Dr. Sugeng Wahyudi, NITTOC Construction Co., Ltd., Japan
Dr. Divyesh Rohit, Kyushu University, Japan

Members

Prof. Fauziah Ahmad, Universiti Sains Malaysia, Malaysia
Mr. Tadashi Akagawa, Japan Foundation Engineering Co., Ltd., Japan
Prof. Guangqi Chen, Kyushu University, Japan
Dr. Takuro Fujikawa, Fukuoka University, Japan
Mr. Takashi Fujishiro, Geo-Disaster Prevention Institute, Japan
Prof. Yasuhide Fukumoto, Kyushu University, Japan
Mr. Hideo Furuichi, Giken Corporation, Japan
Dr. Zentaro Furukawa, Kyushu University, Japan
Dr. Stuart Kenneth Haigh, University of Cambridge, UK
Mr. Shinichiro Ishibashi, Nihon Chiken Co., Ltd., Japan
Dr. Tomohiro Ishizawa, National Research Institute for Earth Science and Disaster Resilience, Japan
Dr. Yusaku Isobe, IMAGEi Consultant Corporation, Japan
Prof. Haruichi Kanaya, Kyushu University, Japan
Dr. Kazuhiro Kaneda, Takenaka Corporation, Japan
Mr. Hiroaki Kaneko, Japan Foundation Engineering Co., Ltd., Japan
Dr. Yoshifumi Kochi, K's Lab Inc., Japan
Mr. Daisuke Matsumoto, Japan Foundation Engineering Co., Ltd., Japan
Dr. Masanori Murai, Shimizu Corporation, Japan
Prof. Juichi Nakazawa, Prof. Emeritus, National Institute of Technology, Maizuru College, Japan
Dr. Shunsaku Nishie, Chuo Kaihatsu Corporation, Japan
Prof. Masashi Sakai, Prof. Emeritus, Kyushu Sangyo University, Japan
Mr. Taisuke Sasaki, Nihon Chiken Co., Ltd., Japan
Dr. Satoquo Seino, Kyushu University, Japan
Prof. Takayuki Shimaoka, Kyushu University, Japan
Prof. Naoaki Suemasa, Tokyo City University, Japan
Dr. Makoto Takada, Chuo Kaihatsu Corporation, Japan
Prof. Takeshi Tsuji, Kyushu University, Japan

Dr. Naoto Watanabe, KFC Ltd., Japan
Dr. Sugeng Wahyudi, NITTOC Construction Co., Ltd., Japan
Mr. Shigeo Yamamoto, Chuo Kaihatsu Corporation, Japan
Dr. Norimasa Yoshimoto, Yamaguchi University, Japan

International Advisory Committee

Advisory Board

Prof. Juichiro Akiyama, Prof. Emeritus, Kyushu Institute of Technology, Japan
Prof. Akira Asaoka, Prof. Emeritus, Nagoya University, Japan
Prof. Ramanathan Ayothiraman, Indian Institute of Technology Delhi, India
Prof. G. L. Sivakumar Babu, Indian Institute of Science Bangalore, India
Prof. Richard J. Bathurst, Royal Military College, Canada
Prof. Adimoolam Boominathan, Indian Institute of Technology Madras, India
Prof. Ross W. Boulanger, University of California Davis, USA
Prof. Deepankar Choudhury, Indian Institute of Technology Bombay, India
Prof. Misko Cubrinovski, University of Canterbury, New Zealand
Mr. Shunta Dozono, Ministry of Land, Infrastructure, Transport and Tourism, Japan
Prof. Tuncer B. Edil, University of Wisconsin-Madison, USA
Prof. Ahmed W. Elgamal, University of California San Diego, USA
Prof. Shailesh R. Gandhi, Sardar Vallabhbhai National Institute of Technology Surat, India
Prof. Noriaki Hashimoto, Kyushu University, Japan
Prof. Yasuhiro Hayashi, Kyushu Sangyo University, Japan
Prof. Takenori Hino, Saga University, Japan
Prof. Akihiko Hirooka, Kyushu Institute of Technology, Japan
Prof. Masayuki Hyodo, Prof. Emeritus, Yamaguchi University, Japan
Mr. Masahiro Inada, Ministry of Land, Infrastructure, Transport and Tourism, Japan
Prof. Buddhima Indraratna, University of Technology Sydney, Australia
Prof. Kenji Ishihara, Chuo University, Japan
Prof. Sangseom Jeong, Yonsei University, South Korea
Prof. Masashi Kamon, Prof. Emeritus, Kyoto University, Japan
Prof. Takeshi Katsumi, Kyoto University, Japan
Prof. Hiroyoshi Kiku, Kanto Gakuin University, Japan
Prof. Yoshiaki Kikuchi, Tokyo University of Science, Japan
Prof. Takeshi Kodaka, Meijo University, Japan
Dr. Eiji Kohama, Port and Airport Research Institute, Japan
Prof. Takaji Kokusho, Prof. Emeritus, Chuo University, Japan
Prof. Junichi Koseki, University of Tokyo, Japan
Prof. Tatsuya Koumoto, Prof. Emeritus, Saga University, Japan
Prof. Osamu Kusakabe, International Press-In Association, Japan

Prof. Johan Lauwereyns, Kyushu University, Japan
Dr. Kyu Tae Lee, Nippon Koei Co., Ltd., Japan
Dr. Wei Feng Lee, Ground Master Construction Co., Ltd., Taiwan
Prof. San Shyan Lin, National Taiwan Ocean University, Taiwan
Prof. Kenichi Maeda, Nagoya Institute of Technology, Japan
Prof. Shunshuke Managi, Kyushu University, Japan
Prof. Masyhur Irsyam, Bandung Institute of Technology, Indonesia
Prof. Taiji Matsuda, Kyushu University, Japan
Prof. Tatsunori Matsumoto, Kanazawa University, Japan
Prof. Yoshiki Mikami, Nagaoka University of Technology, Japan
Prof. Yasuhiro Mitani, Kyushu University, Japan
Prof. Nicola Moraci, Mediterranea University of Reggio Calabria, Italy
Prof. Akira Murakami, Kyoto University, Japan
Mr. Kazuya Murayama, Ministry of Land, Infrastructure, Transport and Tourism, Japan
Prof. Hideo Nagase, Kyushu Institute of Technology, Japan
Mr. Iwao Nakahara, Japan Foundation Engineering Co., Ltd., Japan
Prof. Yukio Nakata, Yamaguchi University, Japan
Prof. Charles Wang-wai Ng, Hong Kong University of Science and Technology, Hong Kong
Prof. Hidetoshi Ochiai, Prof. Emeritus, Kyushu University, Japan
Prof. Satoru Ohtsuka, Nagaoka University of Technology, Japan
Prof. Yasuhito Osanai, Kyushu University, Japan
Prof. Jun Otani, Kumamoto University, Japan
Prof. Aleksandr Petriaev, St. Petersburg State Transport University, Russia
Prof. Kyriazis Pitilakis, Aristotle University of Thessaloniki, Greece
Prof. Karpurapu Rajagopal, Indian Institute of Technology Madras, India
Prof. Krishna R. Reddy, University of Illinois Chicago, USA
Prof. Kenichi Sato, Fukuoka University, Japan
Prof. Eun Chul Shin, Incheon National University, South Korea
Dr. Sanjay Kumar Shukla, Edith Cowan University, Australia
Prof. Francesco Silvestri, University of Naples Federico II, Italy
Prof. Thallak G. Sitharam, Indian Institute of Technology Guwahati, India
Prof. Motoyuki Suzuki, Yamaguchi University, Japan
Mr. Yukihisa Takahashi, Taisei Corporation, Japan
Mr. Hideki Tanigawa, JAFEC USA Inc., USA
Prof. Vikas Thakur, Norwegian University of Science and Technology, Norway
Prof. Ikuo Towhata, Specially Appointed Professor, Kanto Gakuin University, Japan
Prof. Lanmin Wang, Lanzhou Institute of Seismology, China
Prof. Koichiro Watanabe, Kyushu University, Japan
Prof. Kentaro Yamamoto, Nishinippon Institute of Technology, Japan
Prof. Atsushi Yashima, Gifu University, Japan
Prof. Susumu Yasuda, Tokyo Denki University, Japan
Prof. Noriyuki Yasufuku, Kyushu University, Japan

Prof. Koki Zen, Prof. Emeritus, Kyushu University, Japan
Prof. Askar Zhussupbekov, Eurasian National University, Kazakhstan

Members

Prof. Ioannis Anastasopoulos, Swiss Federal Institute of Technology in Zurich, Switzerland
Dr. Panjamani Anbazhagan, Indian Institute of Science Bangalore, India
Prof. Netra Prakash Bhandary, Ehime University, Japan
Dr. Gabriele Chiaro, University of Canterbury, New Zealand
Dr. Rajan Kumar Dahal, Tribhuvan University, Nepal
Dr. Kiyoshi Fukutake, Shimizu Corporation, Japan
Prof. Chandan Ghosh, National Institute of Disaster Management, India
Prof. Seyed Mohsen Haeri, Sharif University of Technology, Iran
Dr. Padavala Hari Krishna, National Institute of Technology Warangal, India
Prof. Manzul Kumar Hazarika, Asian Institute of Technology, Thailand
Prof. Shinya Inazumi, Shibaura Institute of Technology, Japan
Dr. Kiyonobu Kasama, Kyushu University, Japan
Dr. Suman Manandhar, Tribhuvan University, Nepal
Dr. John McDougall, Edinburgh Napier University, UK
Dr. Anil Kumar Mishra, Indian Institute of Technology Guwahati, India
Prof. Adapa Murali Krishna, Indian Institute of Technology Tirupati, India
Prof. Mitsu Okamura, Ehime University, Japan
Prof. Kiyoshi Omine, Nagasaki University, Japan
Dr. Rolando P. Orense, University of Auckland, New Zealand
Prof. Devendra Narain Singh, Indian Institute of Technology Bombay, India
Dr. Doni Prakasa Eka Putra, Gadjah Mada University, Indonesia
Dr. Hing-Ho Tsang, Swinburne University of Technology, Australia
Dr. Bui Trong Vinh, Ho Chi Minh City University of Technology, Vietnam
Dr. Jayan S. Vinod, University of Wollongong, Australia
Prof. Chungsik Yoo, Sungkyunkwan University, Korea

Preface

Sustainable Geo-Technologies for Climate Change Adaptation is a compilation of peer-reviewed papers of the plenary lectures, keynote lectures, special lectures and young researcher's special lectures delivered in the 1st International Symposium on Construction Resources for Environmentally Sustainable Technologies (**CREST 2022**) organized by Kyushu University, Fukuoka, Japan. It was co-organized by the University of Cambridge (United Kingdom) and the International Society for Soil Mechanics and Geotechnical Engineering (ISSMGE). It was supported by the Ministry of Land, Infrastructure, Transport and Tourism (MLIT), Japan, Fukuoka Prefecture, Fukuoka City, the Embassy of India in Japan, UN-Habitat in Fukuoka, the International Press-In Association (IPA), the Kyushu Branch of the Japanese Geotechnical Society (JGS), Global and Local Environment Co-creation Institute (GLEC), Ibaraki University and the Japan Federation of Construction Contractors.

The main purpose of this symposium was to disseminate information and exchange ideas on issues related to natural and man-made disasters and to arrive at solutions through the use of alternative resources, towards building a sustainable and resilient society from the geotechnical perspective. The symposium focused on the sustainability, promotion of new ideas and innovations in design, construction and maintenance of geotechnical structures with the aim of contributing towards climate change adaptation and disaster resiliency to meet the Sustainable Development Goals (SDGs) of the UN. The symposium was successful in bringing together scientists, researchers, engineers and policymakers throughout the world under one umbrella for debate and discussion on those issues.

Sustainable Geo-Technologies for Climate Change Adaptation contains the latest research, information, technological advancement, practical challenges encountered and the solutions adopted in the field of geotechnical engineering for sustainable infrastructure towards climate change adaptation. The volume comprises of 16 contributions, which were delivered during the symposium by invited speakers from all over the world. All the manuscripts were thoroughly reviewed by at least two reviewers selected from an international panel of experts.

The publication of *Sustainable Geo-Technologies for Climate Change Adaptation* has been possible through the sustained efforts of the staff of the Research Group of Adaptation to Global Geo-Disaster and Environment (Geo-disaster Prevention Engineering Laboratory), Kyushu University, Japan. The editors express their sincere thanks to all the members of the research group. Special thanks go to Dr. Divyesh Rohit, Co-Secretary of the CREST 2020 technical committee, for his tireless efforts and dedication, which were instrumental in the timely publication of the book. The editors also would like to express their sincere gratitude to all the reviewers for their time and effort to review the manuscripts and improve their contents.

The editors hope that students, researchers, professionals and policymakers will find the contents of this book useful. The editors believe that the knowledge contained in this book will contribute towards achieving the SDGs set forth by the UN in the years to come.

Fukuoka, Japan	Hemanta Hazarika
Cambridge, United Kingdom	Stuart Kenneth Haigh
Fukuoka, Japan	Haruichi Kanaya
Surathkal, India	Babloo Chaudhary
Yamaguchi, Japan	Yoshifumi Kochi
Tokyo, Japan	Masanori Murai
Tokyo, Japan	Sugeng Wahyudi
Kitakyushu, Japan	Takashi Fujishiro

Acknowledgments

Financial Supports

The chairman of the 1st International Symposium on **C**onstruction **R**esources for **E**nvironmentally **S**ustainable **T**echnologies (**CREST 2020**) and the editors of this book gratefully acknowledge the financial support provided by Kyushu University under two special projects (Progress 100 and SHARE-Q) and JAFEC USA Inc., California, USA. The editors also would like to acknowledge all our sponsors (Diamond, Gold, Silver and Bronze), without which holding the symposium and this publication would not have been possible.

Panel of Reviewers

The manuscript for each chapter included in this book was carefully reviewed for the quality and clarity of technical contents by at least two members of the review panel, consisting of the following international experts. The editors wish to express their sincere gratitude to all the reviewers for their valuable time and efforts.

Prof. Guangqi Chen, Kyushu University, Japan
Dr. Awdhesh Kumar Choudhary, National Institute of Technology Jamshedpur, India
Prof. Sujit Kumar Dash, Indian Institute of Technology Kharagpur, India
Dr. Takuro Fujikawa, Fukuoka University, Japan
Prof. Chandan Ghosh, National Institute of Disaster Management, India
Dr. Yi He, Southwest Jiaotong University, China
Dr. Arvind Kumar Jha, Indian Institute of Technology Patna, India
Dr. Vijay Kumar, Motilal Nehru National Institute of Technology Allahabad, India
Dr. Wei Feng Lee, Ground Master Construction Co., Ltd., Taiwan
Prof. San-Shyan Lin, National Taiwan Ocean University, Taiwan
Dr. Suman Manandhar, Tribhuvan University, Nepal

Prof. Kasinathan Muthukkumaran, National Institute of Technology Trichy, India
Dr. Suchit Kumar Patel, Central University of Jharkhand, India
Dr. Dhiraj Raj, Malaviya National Institute of Technology Jaipur, India
Dr. Anil Kumar Sharma, National Institute of Technology Patna, India
Prof. Naoaki Suemasa, Tokyo City University, Japan
Mr. Vamsi Nagaraju Thotakura, Sagi Ramakrishnam Raju Engineering College, India
Prof. Lanmin Wang, Lanzhou Institute of Seismology, China
Dr. Lin Wang, Chuo Kaihatsu Corporation, Japan
Dr. Norimasa Yoshimoto, Yamaguchi University, Japan
Prof. Feng Zhang, Nagoya Institute of Technology, Japan

Contents

Part I Landslides and Slope Failures

1. **Early Warning Practice for Shallow Landslides in Norway and Physical Modelling Strategies Supported by IoT-Based Monitoring** 3
   Emir Ahmet Oguz, Cristian Godoy Leiva, Ivan Depina, and Vikas Thakur

2. **Liquefaction-Induced Flow Failure of Gentle Slopes of Fines—Containing Loose Sands by Case Histories and Laboratory Tests** 17
   Takaji Kokusho, Hemanta Hazarika, Tomohiro Ishizawa, Shin-ichiro Ishibashi, and Katsuya Ogo

3. **A Regional-Scale Analysis Based on a Combined Method for Rainfall-Induced Landslides and Debris Flows** 35
   Sangseom Jeong and Moonhyun Hong

4. **Views on Recent Rainfall-Induced Slope Disasters and Floods** 45
   Ikuo Towhata

5. **Appropriate Technology for Landslide and Debris Flow Mitigation in Thailand** 81
   Suttisak Soralump and Shraddha Dhungana

6. **Slope Creep Instability in Krajang Lor Village, Magelang Regency, Central Java, Indonesia: Inducement and Developmental Prediction** 99
   Tran Thi Thanh Thuy

7. **Application and Feedback Analysis of the Freeway Slope Maintenance Management System in Taiwan** 123
   San-Shyan Lin, Wen-I Wu, Tsai-Ming Yu, Chia-Yun Wei, Lee-Ping Shi, and Jen-Cheng Liao

Part II Characterization of Geo-Materials

8 Chemical and Mechanical Properties of Geopolymers Made of Industrial By-Products Such as Fly Ash, Steel Slags and Garbage Melting Furnace Slags 135
Tatsuya Koumoto

9 Characteristics of Re-liquefaction Behaviors of the Typical Soils in the Aso Area of Kumamoto, Japan 147
Guojun Liu, Noriyuki Yasufuku, Ryohei Ishikura, and Qiang Liu

Part III Sustainable Development for Infrastructures

10 Sustainable Transport Infrastructure Adopting Energy-Absorbing Waste Materials 159
Buddhima Indraratna, Yujie Qi, and Trung Ngo

11 Life Cycle Sustainability Assessment: A Tool for Civil Engineering Research Prioritization and Project Decision Making 175
Alena J. Raymond, Jason T. DeJong, and Alissa Kendall

12 Role of the Indonesian Society for Geotechnical Engineering in the Development of Sustainable Earthquake-Resilience Infrastructure in the Recent Years 185
Arifan Jaya Syahbana, Masyhur Irsyam,
Delfebriyadi Delfebriyadi, Mahdi Ibrahim Tanjung,
Rena Misliniyati, Mohamad Ridwan, Fahmi Aldiamar,
Nuraini Rahma Hanifa, Arifin Beddu, and Agus Himawan

Part IV Adaptation to Climate Change-Induced Hazards

13 Climate Change-Induced Geotechnical Hazards in Asia: Impacts, Assessments, and Responses 197
Kazuya Yasuhara and Dennes T. Bergado

14 Effect of Vessel Waves on Riverbank Erosion: A Case Study of Mekong Riverbanks, Southern Vietnam 225
Tran The Le Dien, Huynh Trung Tin, Bui Trong Vinh,
Trang Nguyen Dang Khoa, and Ta Duc Thinh

15 Sustainability and Disaster Mitigation Through Cascaded Recycling of Waste Tires—Climate Change Adaptation from Geotechnical Perspectives 239
Hemanta Hazarika, Yutao Hu, Chunrui Hao,
Gopal Santana Phani Madabhushi, Stuart Kenneth Haigh,
and Yusaku Isobe

Editors and Contributors

About the Editors

Hemanta Hazarika is Professor at the Graduate School of Engineering and Department of Interdisciplinary Science and Innovation, Kyushu University, Japan. His research activities include disaster prevention and mitigation, soil-structure interaction, stability of soil-structures during earthquakes and tsunami, ground improvement, application of recycled waste and lightweight materials in construction, stability of cut slopes, and landslides and countermeasures to mitigate landslides disasters. He has published over 350 technical papers in various international journals, conferences, workshops and symposia to date. Prof. Hazarika has several years of experience in teaching, research as well as geotechnical practice and consulting both within and outside Japan.

Stuart Kenneth Haigh is Professor in Geotechnical Engineering at Cambridge University, UK and a Fellow of Trinity College, UK. His research interests involve physical and numerical modelling in areas including earthquake engineering, retaining wall and pile behaviour. He has published over 70 papers in various international journals and will be the 2021 Geotechnique Lecturer.

Haruichi Kanaya is Professor in the Graduate School of Information Science and Electrical Engineering, Kyushu University, Japan. In 1998, he was a visiting scholar in the Massachusetts Institute of Technology (MIT), USA. He is a senior member of IEEE, USA and IEICE, Japan. His research areas include superconductors, RF CMOS LSIs, antennas and energy harvesting systems.

Babloo Chaudhary is Assistant Professor in the Department of Civil Engineering, National Institute of Technology Karnataka (NITK) Surathkal, India. His research areas include geo-disaster prevention and mitigation, dynamic soil-structures interaction, ground improvement, coastal geotechnics, energy geotechnics. He has more than 50 technical papers in various international journals, international and national conferences, workshops and symposia.

Yoshifumi Kochi is currently a Representative Director of K's Lab. Inc., Yamaguchi, Japan. He is a specialist in landslides disasters and their mitigations. Dr. Kochi has 17 years of experience in teaching engineering ethics to undergraduate students. He has been involved in geotechnical practice and consulting both within and outside Japan for more than forty years.

Editors and Contributors   xix

Masanori Murai is Senior Engineer in the Division of Civil Engineering Technology at Shimizu Corporation, Tokyo, Japan. He received his Ph.D. in Geology from Kochi University, Japan in 2007. As a professional civil engineer, he has participated in many projects related to infrastructure development and landslide mitigation.

Sugeng Wahyudi is Manager in Engineering and Development Division at NITTOC Construction Co., Ltd., Tokyo, Japan. He obtained doctoral degree in Engineering from Kyushu University in 2011. Before joining NITTOC in 2020, he worked at National Institute of Advanced Industrial Science and Technology (AIST), Japan and Kyushu University, Japan. His research interests include nanobubble technology and AI to develop advanced and smart solutions for construction site.

Takashi Fujishiro is the representative of the Ground Disaster Prevention Research Institute, Kitakyushu, Japan, a company which he started in August 2019. He specializes in geology and has several years of experiences in surveys and design of slope disaster countermeasures, and road planning and maintenance.

Contributors

Aldiamar Fahmi Faculty of Civil and Environmental Engineering, Bandung Institute of Technology, Bandung, Indonesia;
Institute of Road Engineering, Ministry of Public Works and Housing, Bandung, Indonesia

Beddu Arifin Faculty of Civil Engineering, Tadulako University, Palu, Indonesia

Bergado Dennes Asian Institute of Technology, Bangkok, Thailand

DeJong Jason T. University of California, Davis, CA, USA

Delfebriyadi Delfebriyadi Faculty of Civil and Environmental Engineering, Bandung Institute of Technology, Bandung, Indonesia;
Research Center for Disaster Mitigation, Bandung Institute of Technology, Bandung, Indonesia

Depina Ivan Department of Rock and Geotechnical Engineering, SINTEF, Trondheim, Norway;
Faculty of Civil Engineering, Architecture and Geodesy, University of Split, Split, Croatia

Dhungana Shraddha Geotechnical Engineering Research and Development Center, Kasetsart University, Bangkok, Thailand

Haigh Stuart Kenneth University of Cambridge, Cambridge, UK

Hanifa Nuraini Rahma Research Center for Disaster Mitigation, Bandung Institute of Technology, Bandung, Indonesia;
Center for Earthquake Science and Technology, Bandung Institute of Technology, Bandung, Indonesia

Hao Chunrui Kyushu University, Fukuoka, Japan

Hazarika Hemanta Geo-Disaster Prevention Engineering Research Laboratory, Kyushu University, Fukuoka, Japan

Himawan Agus Faculty of Civil and Environmental Engineering, Bandung Institute of Technology, Bandung, Indonesia

Hong Moonhyun Yonsei University, Seoul, Republic of Korea

Hu Yutao Kyushu University, Fukuoka, Japan

Indraratna Buddhima Director of Transport Research Centre (TRC) and Founding Director of Australian Research Council's Industrial Transformation Training Centre for Advanced Technologies in Rail Track Infrastructure (ITTC-Rail), Faculty of Engineering and Information Technology, University of Technology Sydney, Sydney, NSW, Australia

Irsyam Masyhur Faculty of Civil and Environmental Engineering, Bandung Institute of Technology, Bandung, Indonesia;
Research Center for Disaster Mitigation, Bandung Institute of Technology, Bandung, Indonesia

Ishibashi Shin-ichiro Nihon Chiken Company Ltd, Fukuoka, Japan

Ishikura Ryohei Kyushu University, Fukuoka, Japan

Ishizawa Tomohiro National Research Institute for Earth Science and Disaster Resilience, Tsukuba, Japan

Isobe Yusaku IMAGEi Consultant Corporation, Tokyo, Japan

Jeong Sangseom Yonsei University, Seoul, Republic of Korea

Kendall Alissa University of California, Davis, CA, USA

Khoa Trang Nguyen Dang Ho Chi Minh City University of Technology – VNU-HCMC, Ho Chi Minh City, Vietnam

Kokusho Takaji Chuo University, Tokyo, Japan

Koumoto Tatsuya Saga University, Saga, Japan

Le Dien Tran The Ho Chi Minh City University of Technology – VNU-HCMC, Ho Chi Minh City, Vietnam;
Rentop Corp, Ho Chi Minh City, Vietnam

Leiva Cristian Godoy Department of Rock and Geotechnical Engineering, SINTEF, Trondheim, Norway

Liao Jen-Cheng Taiwan Construction Research Institute, New Taipei City, Taiwan

Lin San-Shyan National Taiwan Ocean University, Keelung, Taiwan

Liu Guojun Changshu Institute of Technology, Changshu, Jiangsu, China

Liu Qiang Shandong University of Science and Technology, Qingdao, Shandong, China

Madabhushi Gopal Santana Phani University of Cambridge, Cambridge, UK

Misliniyati Rena Faculty of Civil and Environmental Engineering, Bandung Institute of Technology, Bandung, Indonesia;
Research Center for Disaster Mitigation, Bandung Institute of Technology, Bandung, Indonesia

Ngo Trung Transport Research Centre, Faculty of Engineering and Information Technology, University of Technology Sydney, Sydney, NSW, Australia

Ogo Katsuya Nippon Koei Company Ltd., Osaka, Japan

Oguz Emir Ahmet Norwegian University of Science and Technology, Trondheim, Norway

Qi Yujie Transport Research Centre, Faculty of Engineering and Information Technology, University of Technology Sydney, Sydney, NSW, Australia

Raymond Alena J. University of California, Davis, CA, USA

Ridwan Mohamad Research Center for Disaster Mitigation, Bandung Institute of Technology, Bandung, Indonesia;
Research Institute for Housing and Human Settlements, Ministry of Public Works and Housing, Bandung, Indonesia

Shi Lee-Ping Taiwan Construction Research Institute, New Taipei City, Taiwan

Soralump Suttisak Geotechnical Engineering Research and Development Center, Kasetsart University, Bangkok, Thailand

Syahbana Arifan Jaya Faculty of Civil and Environmental Engineering, Bandung Institute of Technology, Bandung, Indonesia;
Research Center for Disaster Mitigation, Bandung Institute of Technology, Bandung, Indonesia

Tanjung Mahdi Ibrahim Faculty of Civil and Environmental Engineering, Bandung Institute of Technology, Bandung, Indonesia;
Research Center for Disaster Mitigation, Bandung Institute of Technology, Bandung, Indonesia

Thakur Vikas Norwegian University of Science and Technology, Trondheim, Norway

Thinh Ta Duc Hanoi University of Mining and Geology, Ha Noi, Vietnam

Thuy Tran Thi Thanh Project Management Unit of Construction Investment, People's Committee of District 12, Ho Chi Minh City, Viet Nam

Tin Huynh Trung Ho Chi Minh City University of Technology – VNU-HCMC, Ho Chi Minh City, Vietnam

Towhata Ikuo Department of Civil Engineering, Kanto Gakuin University, Yokohama, Japan

Vinh Bui Trong Ho Chi Minh City University of Technology – VNU-HCMC, Ho Chi Minh City, Vietnam

Wei Chia-Yun Taiwan Area National Freeway Bureau, New Taipei City, Taiwan

Wu Wen-I Taiwan Area National Freeway Bureau, New Taipei City, Taiwan

Yasufuku Noriyuki Kyushu University, Fukuoka, Japan

Yasuhara Kazuya GLEC, Ibaraki University, Ibaraki, Japan

Yu Tsai-Ming Taiwan Area National Freeway Bureau, New Taipei City, Taiwan

Part I
Landslides and Slope Failures

Chapter 1
Early Warning Practice for Shallow Landslides in Norway and Physical Modelling Strategies Supported by IoT-Based Monitoring

Emir Ahmet Oguz, Cristian Godoy Leiva, Ivan Depina, and Vikas Thakur

1.1 Introduction

Water-triggered shallow landslides are one of the major hazards in Norway. They are initiated by extreme events of rainfall, snowmelt or a combination of both. Such events can trigger landslides through several mechanisms including increasing soil water content and increasing soil weight, decreasing suction, erosion and artesian pressure. In Norway, steep slopes and modified landscapes, such as excavations and fillings for roads and railways, are highly prone to this kind of landslide.

The studies on documenting fatalities by the Norwegian Water Resources and Energy Directorate (NVE) show that 45 people died in Norway in the period between 1995 and 2019 due to soil landslides categorized as rockfalls and small rock avalanches (18), rock slides (5), slush flow (7), debris flow and debris avalanche (7) and clay slide (8). World Bank reported that 3.7×10^6 km^2 of the land surface is landslide-prone and that 300 million people are exposed to landslide risk (Dilley et al. 2005). In the study by Petley (Petley 2012), 2620 deadly landslides were documented worldwide from 2004 to 2010, causing 32,322 fatalities which is the highest reported value so far in reports. Similarly, studies on economic losses caused by landslides show that landslides cause damage to infrastructures such as roads, railways, pipelines, structures, embankments, buildings and other built environments.

E. A. Oguz (✉) · V. Thakur
Norwegian University of Science and Technology, Trondheim, Norway
e-mail: emir.a.oguz@ntnu.no

C. G. Leiva · I. Depina
Department of Rock and Geotechnical Engineering, SINTEF, Trondheim, Norway

I. Depina
Faculty of Civil Engineering, Architecture and Geodesy, University of Split, Split, Croatia

© The Author(s), under exclusive license to Springer Nature Singapore Pte Ltd. 2023
H. Hazarika et al. (eds.), *Sustainable Geo-Technologies for Climate Change Adaptation*, Springer Transactions in Civil and Environmental Engineering,
https://doi.org/10.1007/978-981-19-4074-3_1

The total global annual losses caused by landslides are reported to be about 18 billion €. This corresponds to 17% of the annual average global natural disaster losses of 110 billion € (Munich Re 2015; Haque et al. 2016). In Norway, claims to the insurance companies due to the floods and landslides reached approximately 275 million € per year only in the south-eastern part, and the overall insurance payments show an increasing trend (Krøgli et al. 2018).

These landslide events are likely to become more common as the climate trends for Norway display increasing temperatures and rainfall amounts, respectively, by 4.6 °C and 30% by the end of the century (Klima-og miljødepartementet (KLD): Klimatilpasning i Norge 2013). Consequently, the frequency of intensive landslide triggering rainfall events is expected to increase with the projected climate change. Besides the increasing landslide hazards, the population is expanding towards landslide-prone areas, which will increase the risk associated with the landslides. The increasing risk of landslides in society motivates the development of efficient landslide risk management strategies to mitigate the potential consequences.

This study provides an overview of the early warning practice for water-triggered landslides in Norway on a regional to national scale by explaining the components of the EWS in Sect. 1.2. A brief review of the physical-based models and their capacity to improve the existing EWS with a case study is provided in Sects. 1.3 and 1.4, respectively. In Sect. 1.4, additionally, a brief overview of IoT-based monitoring strategy and its capacity are discussed.

1.2 The Norwegian Landslide Forecasting and Warning Service

NVE was given the responsibility of establishing a rainfall-induced landslide forecasting service in 2009. The official launch of 'the Norwegian Landslide Forecasting and Warning Service' was accomplished in 2013. The system was developed by joint initiatives including NVE, the Norwegian Meteorological Institute (MET), and the Norwegian Public Road Administration (NPRA). The main components of the Norwegian Landslide Forecasting and Warning Service, which will be called "the service" hereafter, are provided in the following section. In addition, validation of the service performance will be discussed afterwards.

1.2.1 Components

A robust EWS necessitates involving reliable meteorological, hydrological and geotechnical models to predict landslides and issue warnings. In addition to these reliable models, meteorological and hydrological networks to collect data, comprehensive landslide databases and a good communication system are also vital components of a robust EWS. The service components (Krøgli et al. 2018; Devoli et al. 2019) including meteorological forecasts, hydrological models, hydrological and meteorological networks, landslide databases, thresholds, susceptibility maps and web tools for decision and communication will be examined.

The service utilizes daily forecasts of both precipitation and temperature. MET is providing the forecasts by using AROMEMetCoOp and EC weather models which are short-term and long-term models, respectively. The resolution of short- and long-term weather models are 2.5 km (for ≈2.5 day forecast) and 9 km (for 9 day forecast), respectively. The hydrological model is based on the hydrological HBV model with a resolution of 1 km^2 and is capable of simulating runoff, soil saturation and soil frost by using temperature and precipitation data. In the model, the temperature and precipitation forecast data are obtained by downscaling the AROME and EC, which introduce additional uncertainties to the system. The model is running four times a day to provide the data. In addition to the HBV model, the one-dimensional soil water and heat flow model (Soilflow model) runs daily in the proximity of areas where groundwater stations are located.

The service utilizes the network of meteorological, hydrological and hydrogeological stations which are operated by MET, NPRA and the Norwegian Rail Administration (Bane NOR). In Norway, there exist more than 400 hydrological stations covering the whole country to measure the discharge in rivers, snow depth and 70 stations to measure groundwater level (operated by NVE). The historic data collected from the stations such as soil moisture, frost and groundwater level are utilized to control the performance of the hydrological models (HBV and Soilflow model) and to correct wrongly estimated parameters.

In the service, the National Landslide Database has vital importance for threshold determination, calibration of the models and evaluation of the service performance. There exists a mass movement database in Norway (controlled by NVE) and it includes more than 65,000 events belonging to different landslide categories with different levels of recording quality. Thresholds in the service are based on the knowledge of the relationship between landslide data and hydro-meteorological parameters which can be predicted. The hydro-meteorological parameters are obtained by utilizing HBV model and the thresholds are obtained statistically by using the National Landslide Database. After investigating the performance of the system with different parameter relationships, the best performance was obtained with the threshold based on the relationship between the relative water supply and the soil water saturation degree. The relative water supply is obtained by simulations of rain and snowmelt and provided as a percentage of the annual average value for a 30-year period. The soil water saturation degree is the ratio of the simulated total water content in soil to the maximum soil water content for a 30-year period, assumed as fully saturated soil. In Fig. 1.1, the natural thresholds for landslide hazards are provided. The three lines, yellow, orange and red, represent minimum, medium and

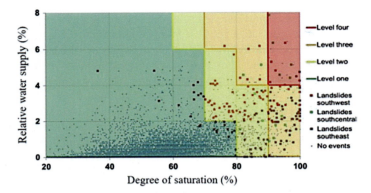

Fig. 1.1 National thresholds for landslide EWS in Norway (www.varsom.no)

maximum national thresholds in Norway, respectively. The existence of some landslide events below the minimum threshold, yellow line, is explained as having poor quality of data registration.

Due to the false alarms resulted from the service, topographical difference and the spatial variability of the climate over the country were considered, and this resulted in obtaining regional thresholds for southern and eastern Norway (Krøgli et al. 2018).

The experience gained through the service in the period of 2011–2012 showed that the rainfall thresholds are giving overestimated danger levels for non-susceptible areas. Then, combining the thresholds of triggering factors (e.g. rainfall) and susceptibility maps was recognized as a way to improve the spatial resolution of thresholds and provide a dynamic hazard assessment (Devoli et al. 2019). In Norway, two main susceptibility maps are in use. The first susceptibility map for landslides in soils is utilized to adjust the threshold values and decrease the overestimated danger levels. The Quaternary geology map, land cover, average yearly rainfall, various runout variables DEM of slope and aspect, etc., are the variables that are combined to get the first susceptibility map. The second map is showing the regions where debris avalanches and debris flows may happen. The source area is determined by the index approach considering several variables, and the flow path and runout are modelled by the Flow-R model. In general, while the first susceptibility map for landslides in soils is used to improve the landslide thresholds, the second susceptibility map for debris is utilized after issued a warning to show possible affected zones.

The researchers developed a tool, called 'xgeo.no', used for decision-making for snow avalanches, landslides and floods. The web portal is in use since 2008 and provides both historic observations, model simulations (hydro-meteorological data), forecasts and real-time data. The data is used by forecasters to determine the warning zonation. The forecast maps are updated four times a day, and forecasts 9 days ahead are also available.

The 'varsom.no' is a web portal that is established to support the dissemination of flood, landslide and snow avalanche warnings. It is open to the public and accessible through smartphones. The assessments, including the forecasts for today, tomorrow

1 Early Warning Practice for Shallow Landslides … 7

and the day after tomorrow, are reported twice a day. The users can choose to be notified about the warning level by an SMS or e-mail. Four importance levels of awareness are used in Norway. That is, red (4), orange (3), yellow (2) and green (1) awareness levels represent conditions from very high danger condition to safe condition.

Additionally, 'regobs.no' web portal was established to register the landslide events in Norway. Forecasters, emergency personnel and normal users can register for the events. All newly registered data are first controlled to check the quality and accuracy before being stored in the mass movement and flood database. In addition to that, the media is also monitored to get information on landslide events. By doing so, improvement and the reliability of the database are achieved.

1.2.2 Validation of the Service

Technical performance, i.e. accuracy of the service is determined in terms of correct alarms, false alarms, missed events and wrong warning levels. The daily assessment is considered as correct if the announced awareness level and real situation are similar. If a lower awareness warning level is issued and some landslide events happen in a region, it is considered as missed events. Lastly, if a day with high levels of awareness does not have the corresponding expected events, either no event or less important events, it is a wrong level or false alarm. Table 1.1 shows the technical performance of the service provided by NVE. It can be seen that the ratio of the events correctly forecasted in the period of 2013–2019 ranges between 92.9 and 98.4% at national scale. Krøgli et al. (2018) reported that the false alarms and missed events are due to the changes in forecasts, some errors in the hydrological models or wrong interpretation of the results. Moreover, the accuracy of the current service can be improved by considering hazard assessment, using reliable weather forecasts and hydrological models or increasing the efficiency of communication.

Table 1.1 Technical performance (%) of the service in Norway

Status\years	2013	2014	2015	2016	2017	2018	2019
Correct	94.2	92.9	97.9	97.8	96.6	98.4	97.9
False alarm	3.3	5.2	1.4	0.8	1.9	0.5	1.0
Missed events	2.2	1.2	0.3	1.1	1.1	0.5	0.8
Wrong level	0.3	0.7	0.4	0.3	0.4	0.5	0.3

1.3 Physical-Based Modelling Strategies

In order to improve the existing EWS in Norway, one way is to employ physical-based models. By doing so, the geotechnical point of view can be also considered in the EWS with the consideration of failure mechanism and soil mechanics. Physical-based models are increasingly utilized in landslide hazard and susceptibility mapping. A wide range of models are available to provide landslide hazard/susceptibility assessment from local (single slope to 10 km^2) to national scales (hundreds to thousands of km^2). These models mostly feature hydrological and geotechnical models, which are responsible for modelling the rainfall infiltration process and slope stability, respectively. That is, the models calculate the pore water pressure values and corresponding safety margins during a rainfall event. Commonly implemented physical-based models include distributed Shallow Landslide Analysis Model (dSLAM), Stability INdex MAPping (SINMAP), Shallow Slope Stability Model (SHALSTAB) and Transient Rainfall Infiltration and Grid-Based Regional Slope Stability (TRIGRS) model (Park et al. 2011).

All aforementioned models are subjected to several limitations arising from the assumptions and approximations in the theory behind them. Due to the complexity of the soil response and high computational effort for large scale, these models mainly focus on shallow landslides with the infinite slope stability method being employed to model the failure mechanism of the soil. The shallow landslides consist of unsaturated zone and saturated zone with a capillary fringe. The majority of the physical-based models do not account for the unsaturated zone of the soil profile. TRIGRS model (Baum et al. 2008) is one of the most commonly used physical-based models as it includes the effect of the unsaturated zone of the soil on the infiltration process and the slope stability assessment. Some amount of water infiltrated from the ground surface is stored in the unsaturated zone. The remaining water that passed the unsaturated zone accumulates on the water table and results in a water table rise. Then, corresponding water pressures (either positive or negative) is used in the slope stability model. In this study, TRIGRS model has been implemented to obtain spatiotemporal safety distribution of shallow water triggered landslides and described in the following section.

1.3.1 Transient Rainfall Infiltration and Grid-Based Regional Slope Stability (TRIGRS)

TRIGRS is a FORTRAN code developed to obtain a spatiotemporal distribution of safety factors for shallow and rainfall-induced landslides. Models for infiltration and subsurface flow (hydrological model), routing of runoff and slope stability have been integrated into the TRIGRS code to examine the response of large areas to excess rainfall events. All the input parameters of TRIGRS model are allowed to vary over a large area to capture the varying nature of these parameters such as a change in initial water depth, slope, precipitation intensity, etc.

The infiltration model is based on the solution of the one-dimensional Richards equation, which describes the movement of water through soil profile in the vertical direction. TRIGRS accounts for both wet initial (saturated) and unsaturated initial conditions. Iverson's linearized solution of Richards equation (Srivastava and Yeh 1991) has been used in the infiltration model for wet initial conditions. The solution consists of long-term (steady, background pressure head) and short-term (transient, time-varying pressure head) for rainfall events with constant intensity. To capture time-varying rainfall events, an extended version of Iverson's solution has been implemented by using the Heaviside step function. For the unsaturated conditions, exponential hydraulic parameter models for hydraulic conductivity and water content (Gardner 1958) are used to linearize the Richard equation and include the unsaturated flow (Srivastava and Yeh 1991).

Like the other physical-based models, TRIGRS utilizes infinite slope stability method in which the pressure head from the infiltration model is used. The general infinite slope stability equation (Eq. 1.1.) is used to calculate the factor of safety, F_S, with depth.

$$F_S(Z, t) = \tan(\varphi')/\tan(\alpha) + \left(c' - \psi(Z, t)\gamma_w \tan(\varphi')\right)/(\gamma_s Z \sin(\alpha)\cos(\alpha)) \tag{1.1}$$

where c' is the effective cohesion, φ' is the effective friction angle, γ_w and γ_s are unit weight of water and soil, respectively, ψ is the water pressure head, Z is the distance in the slope-normal direction, t is the time and α is the slope angle. For the unsaturated zone, Bishop's effective stress parameter, $\chi = (\theta - \theta_r)/(\theta_s - \theta_r)$, is utilized and pressure head is multiplied by χ. F_S calculations are performed on a cell-by-cell basis (i.e. individually for each cell) by using the grid-based cell parameters.

TRIGRS also accounts for the runoff of excess rainfall water due to soil saturation or the exceedance of infiltrability of the soil, simply K_s. When the infiltration capacity of the soil is exceeded or the soil is saturated, excess precipitation is distributed to the adjacent area. For the runoff procedure, the distribution weighting factors are calculated for the adjacent cells. TRIGRS provides different methods to calculate the weighting factors such as D8 method, D-infinity method, uniform distribution method, slope proportion distribution method, etc. For the details of the calculation, the manual (Baum et al. 2008) can be seen.

Being a grid-based model, all pressure head and F_S calculations are performed for each cell individually. Taking into account unsaturated soil zone, allowing water table rise and giving several options to the users to distribute the excess water to the adjacent cells are advantageous parts of TRIGRS model. However, the model includes several assumptions associated with infiltration, runoff routing and slope stability model. The infiltration model assumes one-dimensional infiltration in homogeneous isotropic soil. In case of strong anisotropy or heterogeneity, the results would be not realistic.

In addition, the strong effect of lateral flow is ignored during prolonged rainfall and between rainfall events. Being sensitive to initial conditions, employing simple routing of surface runoff and simple infinite slope stability models are also other disadvantages of the TRIGRS model. The analysis should be performed considering all cons and pros of the model.

Implementing a physical-based model in the service could improve the resolution of EWS and provide more accurate system performance on a local to regional scale. Both topographical changes, spatial variability of climate, geotechnical property variability can be combined into one model and susceptibility maps based on F_S can be obtained. In Sect. 1.4, the TRIGRS model has been implemented in a landslide-prone area in Norway and the results are presented.

1.4 Case Study

1.4.1 Landslide Susceptibility Assessment with TRIGRS

The TRIGRS model has been implemented in a landslide-prone area, covering around 200 km² of the Stjørdalselva river catchment, between Hegra and Meråker. The study area is located in the Trøndelag region of Norway, 45 km north-east of Trondheim (Fig. 1.2a). The average annual precipitation recorded at Hegra and Meråker Stations is 964 mm and 1205 mm, respectively. There exist steep slopes along the Stjørdalselva river with slope angles greater than 30°. In the report of NVE, the study region is reported to have very high landslide susceptibility (Devoli et al. 2019).

Based on the available geological maps (the Geological Survey of Norway, NGU 2019), the materials present in the site are divided into three classes; sedimentary deposits consisting of fluvial, marine, fjord and moraine materials, organic deposits consisting of peat and marshes and rock and block field deposits as shown in Fig. 1.2b. These classes are named sedimentary, organic deposit and rock. The geotechnical

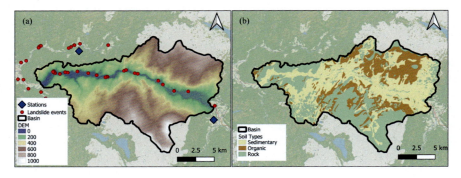

Fig. 1.2 **a** DEM and **b** simplified geological units of the study area, Hegra and Meråker region

Table 1.2 Geotechnical soil parameters of soil classes: organic deposit, rock and sedimentary

Parameter	Organic	Rock	Sedimentary
Cohesion (kPa)	20.0	100.0	5.7
Friction angle (°)	0.0	44	22.2
Sat. permeability (m/s)	1×10^{-6}	1×10^{-6}	1.80×10^{-5}
Diffusivity (m²/s)	4×10^{-5}	5×10^{-6}	1.62×10^{-4}
Unit weight (kN/m³)	15.0	25.0	19.0

properties of the organic deposit layer are directly obtained from the literature (Melchiorre and Frattini 2012). Parameters of the rock layer have been assigned so that it stays stable before and after the rainfall event as no soil-related landslide events are registered on this layer. However, the sedimentary layer has been investigated in detail due to a great proportion of the landslide events being in this zone. The parameters of the sedimentary layer have been obtained by calibration of the model with respect to the registered landslides. Geotechnical parameters of each soil type are given in Table 1.2.

DEM of the study area was obtained from "hoydedata.no" with a resolution of 10 m. Slope angles and direction of runoff were derived from DEM with the same resolution. The thickness of the soil, i.e. depth to bedrock, was calculated for each cell based on the relationship between the groundwater well data obtained from NGU and slope angles. A minimum thickness of 20 cm is assigned in case of correlation resulting in lower thicknesses. Additionally, 0.5 m thickness is assigned to the cells which is reported to be shallow in the NGU geotechnical maps and rock zones which are not potential zones for soil landsliding.

The National Landslide Database of Norway has been investigated for the study area. From the list of recorded movements, only soil-related landslides were extracted. Moreover, some of the data were eliminated dependent on the quality of the record and the date. In the final analysis, only 22 landslide records have been considered to be soil-related landslides with good quality. Eight of them have been used in the calibration process as the remaining events are located in flat areas and cannot be predicted by TRIGRS model. For each landslide event, the corresponding rainfall amount (mm/day) has been obtained from the meteorological stations close to the study area. It is seen that the precipitation amounts range from very low or even zero for a few landside events to high intensity such as 71.7 mm/day, which has been employed in the calibration of the model so that the intensity might trigger all landslide events.

For susceptibility mapping, precipitation with a 100-year period has been utilized. To calculate the amount of probable maximum precipitation with a 100-year return period, the model proposed by Førland and Kristoffersen (Førland and Kristoffersen 1989) (Eq. 1.2) has been utilized.

$$M_T = M_5 \cdot \exp(\lambda \cdot (\ln(T - 0.5) - 1.5)) \qquad (1.2)$$

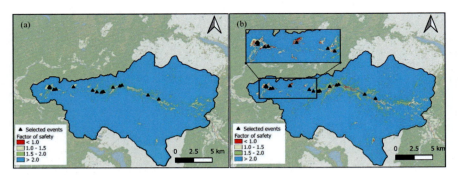

Fig. 1.3 Susceptibility map of Hegra and Meråker region, **a** before and **b** after 24 h precipitation with a 100-year return period

where M_T and M_5 are 24 h rainfall with a T-year and 5-year return period, respectively, T is the return period and λ is a function based on M_5 and can be calculated by $(0.3584 - 0.0473 \cdot \ln(M_5))$, if $25 < M_5 \leq 200$. Precipitation value with a 5-year return period has been obtained as 43.5 mm/day by considering six weather stations around the study area. By using M_5, M_{100} was calculated as 75.9 mm/day.

In this study, the solution for unsaturated flow in TRIGRS model has been employed for the sedimentary zone so that the actual response of the area can be obtained. In addition to the saturated permeability, saturated water content (θ_s) and residual water content (θ_r) with fit parameter (α) were specified as 0.41, 0.05 and 3.5, respectively. In Hegra and Meråker regions, the soil is highly saturated with a water table close to the ground surface during the snow melting and rainy season. In summer, the water table lowers and more proportion of soil becomes unsaturated. In this study, depth of the water table is assumed to be half of the thickness of soil due to the lack of measurements about the actual conditions.

The results of the analysis are shown in Fig. 1.3, which presents the susceptibility map of the region before and after 24 h precipitation with a 100-year return period, M_{100}. The results of the physical-based model, TRIGRS show a good agreement between the distribution of zones prone to failure with the landslide events. The results seem to overpredict the failure compared with observations. This might be attributed to the simplifications and assumptions included in the physical-based model. In addition, the effect of the vegetation and slope curvature (convex or concave) are not considered in the model. In this study, a deterministic approach has been employed. That is, the soil parameters have only one value and is constant over the site. The results of the model can be improved by including the variability of the soil parameters at point and through distance. Besides, the vegetation cover can be considered to see how the results are affected.

1.4.2 Strategy to Adapt IoT-Based Monitoring in Physical-Based Modelling

One of the main challenges in an accurate water-triggered landslide prediction model is the uncertainties commonly associated with geotechnical (e.g. strength values), hydrological (e.g. groundwater levels) and meteorological (e.g. precipitation) parameters controlling the stability of the slope. By adopting IoT-based monitoring, it is possible to collect data on triggering variables and current hydrological and meteorological conditions at the site. The collected data, then, can be used in a probabilistic framework, such as Bayesian updating (Depina et al. 2017, 2020; Straub and Papaioannou 2014) to model the uncertainties related to the model parameters and the uncertainties can be reduced (Depina et al. 2019).

As a landslide risk management strategy, monitoring and EWS based on sensor-based solutions and real-time monitoring have been increasingly employed. With the studies on sensor-based monitoring and EWSs (Baum et al. 2005; Smith et al. 2009; Pecoraro et al. 2018), it has been shown that a more consistent and reliable hazard assessment based on collected data on the triggering variables, and reduction of consequences with timely warnings to protect the elements under risk can be achieved. The IoT is a new and powerful concept in monitoring and EWS and can be simply defined as a system (network) of connected devices that can transfer and exchange information without requiring human intervention. IoT-based systems have the potential to provide cost-effective, flexible and scalable monitoring and EWS, and can provide significant automatization of the risk management by utilizing advanced data analysis, statistical learning algorithms and machine learning algorithms integrated with the landslide prediction models and EWSs. In the following paragraph, a brief overview of IoT system will be provided.

A general IoT system consists of four dependent layers, namely perception, network, middleware and application layers. The perception layer, as the first layer, is a network of IoT devices responsible for sensing, actuating, controlling and monitoring activities (Ray 2018). Simply, the main function of the perception layer is data collection through IoT devices which gather the data from sensors and exchange the data with connected devices. The second layer, network layer, collects the data from the perception layer and transfers to the next layer. In the network layer, significant improvements can be achieved by adopting networking solutions supporting low-power wide-area networks (LPWANs). LPWANs can transfer the data on longer distances and provide power and cost efficiency compared to conventional solutions such as Wi-Fi or Bluetooth. The next layer is the middleware layer, which might have a wide range of functionality depending on the implementation. These functionalities might be data collection, storage, filtering, transformation and data processing by using advanced statistical methods or machine learning. Afterwards, the end-user can access the data through the last layer, the application layer in which the data can be integrated within the monitoring and EWSs. That is, the application layer may support the EWS with the integration of landslide prediction models and the collected data. More details can be seen in Oguz et al. (2019).

1.5 Conclusions

The study presented the current state of the art of early warning practice for shallow landslides in Norway on a regional to national scale. Additionally, physical-based models were briefly described and TRIGRS model was implemented on a landslide-prone area to evaluate the capacity of the model to improve the EWS. The model showed promising results and potential to innovate the existing system in Norway. Furthermore, a brief overview of IoT-based systems was provided by describing their components.

The results of TIGRS implementation are in good agreement with the registered landslide events. This indicates that the model is suitable for susceptibility mapping and has the potential to improve existing EWS in Norway by considering geotechnical, hydrological and meteorological aspects of the region and providing high resolution (10 m). More comprehensive and robust EWS can be achieved by employing physical-based models supported by IoT-based monitoring of the region.

Acknowledgements The authors gratefully acknowledge the support from Dr. Graziella Devoli (NVE), the Research Council of Norway, the partners through KlimaDigital (www.klimadigital.no), and a strategic initiative by NTNU related to Resilient and Sustainable Water Infrastructure (www.sfiwin.com).

References

Baum RL, McKenna JP, Godt JW, Harp EL, McMullen SR (2005) Hydrologic monitoring of landslide-prone coastal bluffs near edmonds and everett, Washington, pp 2001–2004

Baum RL, Savage WZ, Godt JW (2008) TRIGRS—A fortran program for transient rainfall infiltration and grid-based regional slope-stability analysis, version 2.0: U.S. Geological Survey Open-File Report, 2008-1159

Depina I, Ulmke C, Boumezerane D, Thakur V (2017) Bayesian updating of uncertainties in the stability analysis of natural slopes in sensitive clays. In: Thakur V, L'Heureux J-S, Locat A (eds) Landslides in sensitive clays: from research to implementation. Springer International Publishing, Cham, pp 203–212

Depina I, Oguz EA, Thakur V (2019) Learning about uncertain predictions of rainfall-induced landslides from observed slope performance. In: Ching J, Li DQ, Zhang J (eds) Proceedings of the 7th international symposium on geotechnical safety and risk (ISGSR). Research Publishing, Singapore, Taipei, Taiwan, pp 608–613

Depina I, Oguz EA, Thakur V (2020) Novel Bayesian framework for calibration of spatially distributed physical-based landslide prediction models. Comput Geotech 125:103660. https://doi.org/10.1016/j.compgeo.2020.103660

Devoli G, Bell R, Cepeda J (2019) Susceptibility map at catchment level, to be used in landslide forecasting. Norway (NVE Report)

Dilley M, Chen RS, Deichmann U, Lerner-Lam A, Arnold M, Agwe J, Buys P, Kjekstad O, Lyon B, Yetman G (2005) Natural disaster hotspots: a global risk analysis (English). World Bank, Washington, DC

Førland EJ, Kristoffersen D (1989) Estimation of extreme precipitation in Norway. Nord Hydrol 20:257–276

Gardner WR (1958) Some steady-state solutions of the unsaturated moisture flow equation with application to evaporation from a water table. Soil Sci 85:228–232

Haque U, Blum P, da Silva PF, Andersen P, Pilz J, Chalov SR, Malet JP, Auflič MJ, Andres N, Poyiadji E, Lamas PC, Zhang W, Peshevski I, Pétursson HG, Kurt T, Dobrev N, García-Davalillo JC, Halkia M, Ferri S, Gaprindashvili G, Engström J, Keellings D (2016) Fatal landslides in Europe. Landslides 13:1545–1554

Klima-og miljødepartementet (KLD): Klimatilpasning i Norge (2013)

Krøgli IK, Devoli G, Colleuille H, Boje S, Sund M, Engen IK (2018) The Norwegian forecasting and warning service for rainfall- and snowmelt-induced landslides. Nat Hazards Earth Syst Sci 18:1427–1450. https://doi.org/10.5194/nhess-18-1427-2018

Melchiorre C, Frattini P (2012) Modelling probability of rainfall-induced shallow landslides in a changing climate, Otta. Central Norway Clim Change 113:413–436. https://doi.org/10.1007/s10584-011-0325-0

NGU (2019) NGU Geological Survey of Norway. https://www.ngu.no/

Oguz EA., Robinson K, Depina I, Thakur V (2019) IoT-based strategies for risk management of rainfall-induced landslides: a review. In: Ching J, Li DQ, Zhang J (eds) Proceedings of the 7th international symposium on geotechnical safety and risk (ISGSR). Research Publishing, Singapore, Taipei, Taiwan, pp 728–733

Park DW, Nikhil NV, Lee SR (2013) Landslide and debris flow susceptibility zonation using TRIGRS for the 2011 Seoul landslide event. Nat Hazards Earth Syst Sci 2833–2849. https://doi.org/10.5194/nhess-13-2833-2013

Pecoraro G, Calvello M, Piciullo L (2018) Monitoring strategies for local landslide early warning systems. Landslides. https://doi.org/10.1007/s10346-018-1068-z

Petley D (2012) Global patterns of loss of life from landslides. Geology 40:927–930. https://doi.org/10.1130/G33217.1

Ray PP (2018) A survey on Internet of Things architectures. J King Saud Univ Comput Inf Sci 30:291–319. https://doi.org/10.1016/j.jksuci.2016.10.003

Re Munich (2015) Review of natural catastrophes in 2014: Lower losses from weather extremes and earthquakes. https://www.munichre.com/en/company/media-relations/media-information-and-corporate-news/media-information/2015/2015-01-07-review-of-natural-catastrophes-in-2014-lower-losses-from-weather-extremes-and-earthquakes.html

Smith JB, Godt JW, Baum RL, Coe JA, Burns WJ, Lu N, Morse MM, Sener-Kaya B, Kaya M (2014) Hydrologic monitoring of a landslide-prone hillslope in the Elliott state forest, Southern Coast Range, Oregon, pp 2009–2012

Srivastava R, Yeh T-J (1991) Analytical solutions for one-dimensional, transient infiltration toward the water table in homogeneous and layered soils. Water Resour Res 27:753–762. https://doi.org/10.1029/90WR02772

Straub D, Papaioannou I (2014) Bayesian updating with structural reliability methods. J Eng Mech 141:04014134. https://doi.org/10.1061/(asce)em.1943-7889.0000839

Chapter 2
Liquefaction-Induced Flow Failure of Gentle Slopes of Fines—Containing Loose Sands by Case Histories and Laboratory Tests

Takaji Kokusho, Hemanta Hazarika, Tomohiro Ishizawa, Shin-ichiro Ishibashi, and Katsuya Ogo

2.1 Introduction

Among earthquake-induced liquefaction case histories, cyclic liquefaction failures on the dilative side of SSL have been observed quite often and archived in many documents. In contrast, cases are very limited where liquefaction-induced flow failures are reported during earthquakes on the contractive side of SSL. There exist, however, two unprecedented and very similar case histories during recent earthquakes in Hokkaido (Satozuka in Sapporo city and Tanno-cho in Kitami city), wherein liquefied sand strangely flowed underground in very gentle man-made fill slopes of a few percent gradient, leaving large ground depression behind. In both of them, a large amount of non-plastic fines was involved in fine sands that presumably made the sand highly contractive and flowable on the contractive side of SSL under the initial shear stress of gentle slopes ≈3%.

T. Kokusho (✉)
Chuo University, Tokyo, Japan
e-mail: koktak@ad.email.ne.jp

H. Hazarika
Geo-Disaster Prevention Engineering Research Laboratory, Kyushu University, Fukuoka, Japan

T. Ishizawa
National Research Institute for Earth Science and Disaster Resilience, Tsukuba, Japan

S. Ishibashi
Nihon Chiken Company Ltd, Fukuoka, Japan

K. Ogo
Nippon Koei Company Ltd, Osaka, Japan

© The Author(s), under exclusive license to Springer Nature Singapore Pte Ltd. 2023
H. Hazarika et al. (eds.), *Sustainable Geo-Technologies for Climate Change Adaptation*, Springer Transactions in Civil and Environmental Engineering,
https://doi.org/10.1007/978-981-19-4074-3_2

In the first part of the paper, the two case histories are outlined to summarize their peculiarities in terms of site and soil conditions as well as their liquefaction failure modes. Then, undrained triaxial test results on sand sampled from one of the case history sites are incorporated to investigate the failure mechanism focusing the role of involved non-plastic fines. Furthermore, the mechanism is discussed in the light of a series of torsional simple shear tests separately conducted on reconstituted sands with various fines content and initial shear stress.

2.2 Case Histories of Liquefaction Flow Failures in Gentle Slopes

2.2.1 Residential Landfill in Sapporo During 2018 Hokkaido Iburi-East Earthquake

2018 Hokkaido Iburi-East earthquake caused unprecedented liquefaction damage in residential landfill in Satozuka, Kiyota-ward, Sapporo city about 50 km distant from the epicenter (Fig. 2.1a). That inflicted considerable settlement and tilting on a number of independent houses due to localized ground depressions exceeding about 3 m maximum, much greater than normal liquefaction-induced subsidence (Fig. 2.1b). Quite different from previous liquefaction cases, no fissures and sand boils occurred on the ground surface in and around the depressions. Lateral surface displacements could not be seen despite long-distance flow of liquefied sand underground. Also note that local settlement of building foundations, relative to supporting ground surface, did not occur. Liquefied sands all fluidized and flowed underground

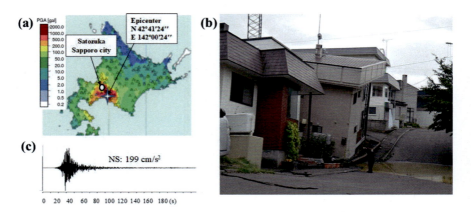

Fig. 2.1 2018 Hokkaido Iburi-East earthquake and liquefied site in Sapporo city 50 km from epicenter (**a**) Photograph of depression (**b**) and Acceleration records at K-NET Hiroshima 8 km from the site (**c**). Modified from NIED (2018)

about 200 m maximum laterally to a margin of the landfill, collectively ejected and went away therefrom. Earthquake motion recorded at about 8 km apart from the site, showed the horizontal peak ground acceleration (PGA) a little lower than 0.2 g, with the predominant frequency 2.5 ~ 5 Hz, and the duration of major motion around 20 s (Fig. 2.1c). Former landscape of the site in 1950s was undulated hills and lowlands of rice field in between. In early 1980s, development of residential land started by cutting the hills (geologically originated from Pleistocene Volcano Shikotsu) and filling the lowlands.

Figure 2.2a shows the present map of damaged area including streets and houses, where filled zones (shaded) are indicated together with a water drainage system that used to be in the old rice field (Sapporo City Office 2018). The thickness of landfill was 5 ~ 9 m. Before the earthquake, the surface gradient of the filled ground along Lines OD shown on the map was 2.6% on average, downslope from D to O. The depression belt around 3.0 m deep maximum occurred along the line OABCD shown in the map about 200 m long almost continuously with the belt width of 20–30 m. Huge volume of liquefied sand underneath the depression belt flowed underground laterally and spout out from Point O. In Fig. 2.2b, the magnitude of ground settlements surveyed after the earthquake is shown stepwise (Sapporo City Office 2018). It is noted that the largest depression belt was almost coincidental with the center of the fill zone where the landfill was thick and traceable down to Point O. The old water drainage system before landfilling was also coincidental with the depression belt as indicated in the figure, leading to a suspicion that this might cause the depression (Yasuda 2018).

In Fig. 2.3a, SPT N-values are plotted versus depth, which were converted from Dynamic Penetration Test (DPT) conducted at P-1, 2 near Point B. The soil was very loose at a few meters depth from the ground surface. Figure 2.3b depicts one of post-earthquake soil investigation results disseminated by Sapporo City Office (2018). Note that a surprisingly loose soil layer of SPT resistance $N \approx 1$ can be observed not

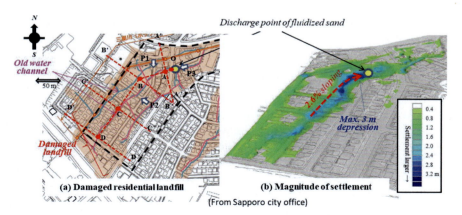

Fig. 2.2 Map of damaged area in Sapporo city with colored fill zone and old water drainage system at the bottom (**a**) and Distribution of ground depressions (**b**) both. Modified from Sapporo City Office

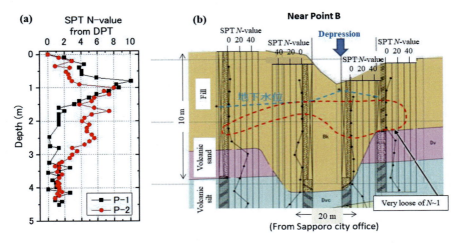

Fig. 2.3 SPT N-value converted from DPT (**a**) and post-earthquake soil investigation results Modified from Sapporo City Office (**b**)

only at the depression but also far from it. This is probably because the landfill was of high fines content ($F_c > 30\%$) and very poorly compacted. Groundwater level in the fill is judged to have been GL. $-2 \sim -3$ m in many points of the fill before the earthquake.

2.2.2 Farm Landfill in Kitami During 2003 Tokachi-Oki Earthquake

A very similar liquefaction case had once occurred in Tanno-cho, Kitami city in Hokkaido, during the 2003 Tokachi-oki earthquake ($M = 8.0$) as reported by Yamashita et al. (2005), Ito et al. (2005) and Tsukamoto et al. (2009), though it did not draw so much attention from general public at that time because it was a rural farmland. The site was 230 km away from the hypocenter of the offshore subduction earthquake Fig. 2.4a, and the maximum acceleration was recorded only 0.055 g at a K-NET station Kitami (about 10 km away from the site) through the duration of major motion was rather long (about one minute), as indicated in Fig. 2.4(b). Analogous to the Sapporo case, the farmland, which was artificially filled with loose volcanic sandy soil and gently inclined (about 3%), liquefied and left considerable local depression behind. However, the ground surface remained intact with no lateral displacement in furrows and no fissures and sand boils in the subsided area as observed in Fig. 2.4c.

As also shown in the air-photograph and plan view of Fig. 2.5a, b, an area, 150 m long and 50 m wide, subsided by 3.5 m maximum and the downslope side was covered by erupted sand from four ejection points and flowed downstream along a ditch (Ito et al. 2005; Yamashita et al. 2005). This again indicates considerable underground

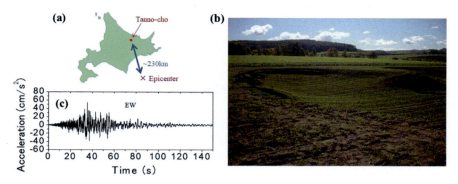

Fig. 2.4 Slope failure in farmland in Kitami during 2003 Tokachi-oki earthquake: **a** Site and epicenter, **b** photo of ground depression, and **c** Acceleration record K-NET Kitami (EW) 10 km from the site

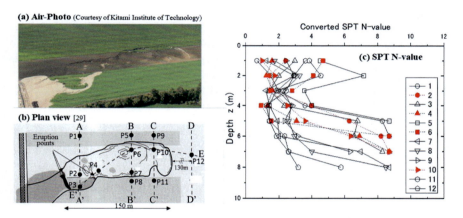

Fig. 2.5 Gentle slope failure in Kitami: **a** Air-photo, **b** soil investigation points, **c** penetration resistance

flowability of liquefied sand of maximum distance 150 m to the ejection points in this case. Unlike the Sapporo case, no drainage pipes were confirmed to have been embedded in the downslope direction according to researchers investigated the site (Yamashita 2019). In situ soil investigation was carried out after the earthquake using the Swedish Weight Sounding (SWS) at 12 points (P1 ~ P12) inside and at the margin of the subsidence zone by Tsukamoto et al. (Tsukamoto et al. 2009). Figure 2.5c shows SPT N-values converted from the SWS test indicating that the sand fill was as loose as $N = 1$ in the loosest portions both inside and outside of the depression. The thickness of soft sandy fill before the earthquake seems to have been variable, 4 ~ 7 m, with the water table 1 ~ 2 m below the ground surface. The fluidization seems to have occurred along an old shallow valley where the fill sand was relatively thick according to the cross-sectional variations of the converted *N*-values (Tsukamoto et al. 2009).

2.2.3 Similarity of the Two Case Histories

Fluidized sands at the two flow failure cases sedimented in the downstream were sampled to investigate their soil properties. The grain-size curves of the two sands are illustrated in Fig. 2.6a, among which Samples **a** and **b** of the Sapporo sand show almost perfect coincidence despite that they were about 50 m apart in the sedimented area. Sample **c** recovered at later time from the same area is still in a good coincidence. This indicates that the Sapporo sand is so uniform that three distant sampling from the wide area of sedimentation made little difference in grain size curves. Furthermore, it is remarkable that the Kitami sand was very similar to the Sapporo sand despite considerable difference in location and event.

In Sapporo, the average grain size is $D_{50} = 0.13$ mm, uniformity coefficient $C_u = 25 \sim 35$, fines content $F_c \approx 35\%$, and the fines were non-plastic (NP). The specific density of soil particle is extraordinarily low, $\rho_s = 2.26 \sim 2.28$ t/m^3, indicating that a large quantity of volcanic pumice (porous and easy to crush) is mixed in the sand. In Kitami, $D_{50} = 0.2$ mm, uniformity coefficient $C_u = 30$, fines content $F_c = 33\%$, and the fines are non-plastic, all of that showed good matching with the Sapporo sand. The specific soil density is low, $\rho_s = 2.47$ t/m^3, again indicating that volcanic pumice is largely mixed in the sand.

Thus, the two very analogous case histories of unprecedented liquefaction-induced ground failure obviously share common features as listed in Table 2.1; (1) they are

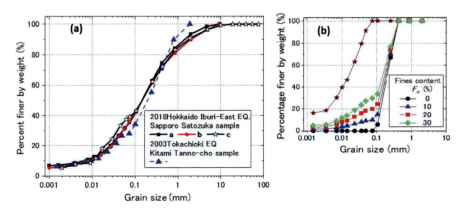

Fig. 2.6 Grain size curves of fluidized sand in Sapporo city during 2018 earthquake compared with that in Kitami city during 2003 earthquake, **a** compared with reconstituted Futtsu beach sand with NP-fines (**b**)

Table 2.1 Similarity of pertinent parameters associated with two flow failures in gentle slopes during two earthquakes

	Sapporo during 2018 EQ	Kitami during 2003 EQ
Surface gradient (%)	2.6	3
SPT N-value	$\approx 1 \sim 5$	$\approx 1 \sim 5$
Ground water table	GL. $-2 \sim -3$ m	GL. $-1 \sim -2$ m
Fines content F_c	$35 \sim 36$	33
Plasticity index I_p	NP	NP
Particle density ρ_s (t/m^3)	$2.26 \sim 2.28$	2.47

gently inclined ($\approx 3\%$) landfills of low density and shallow groundwater, (2) the failures accompany neither ground fissures nor sand boiling unlike normal liquefaction manifestations but considerable depressions of exceeding 3 m because the underground liquefied sand fluidized long-distance downslope and erupted collectively, and (3) the two sands share very similar physical properties. Among them, a lot of NP fines exceeding $F_c = 30\%$ seem to have had the most significant effects on the flowability.

2.3 Undrained Triaxial Tests on Sapporo Sand

In order to know the effect of the non-plastic fines on the soil behavior, a series of undrained triaxial tests have been conducted on the Sapporo sand. Test specimens of 5 cm in diameter and 10 cm in height were reconstituted by the moist-tamping method to make the relative density $D_r =$ around 50–80% after consolidation. Cyclic loading tests and monotonic loading tests were conducted to compare the results between the original sand from in situ and that deprived of fines. The maximum and minimum void ratios were $e_{\max} = 2.231$, $e_{\min} = 1.268$ and $e_{\max} = 2.351$, $e_{\min} = 1.422$, for the sand with and without fines, respectively.

2.3.1 Cyclic Loading Response

Figure 2.7 depicts cyclic stress ratios (*CSR*) versus number of cycles (N_c) obtained under low effective confining stress of $\sigma'_c = 28$ kPa corresponding to shallow depth for the original Sapporo sand of medium density ($D_r = 55\%$) and high densities ($D_r \approx 80\%$; 77 ~ 82%) plotted with close symbols. Note that CRR_{20} (*CSR* for $N_c = 20$) is extremely low (less than 0.1 for $D_r = 55\%$ and nearly 0.2 for $D_r \approx 80\%$ compared to normal sands of those D_r-values. The open star symbols overlaid are test results for the same sand of $D_r = 59 \sim 69\%$ but deprived of all fines (smaller

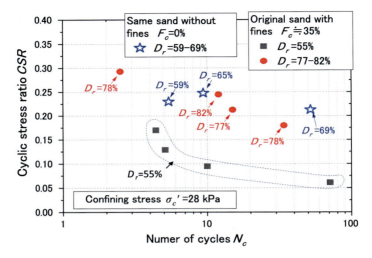

Fig. 2.7 CSR versus N_c for Sapporo sand compared with that deprived of fines

than 75 μm). It is obvious that these plots are positioned higher in comparison of the original sand of $D_r \approx 80\%$.

Figure 2.8a, b illustrates time histories of the original and fine-deprived sands, respectively, in terms deviatoric stress σ_d, pore-pressure and axial strain. Despite the small difference of density ($D_r = 55$ and 59%), the original sand with $F_c \approx 35\%$ tends to behave as if very low-density sand, quite different from that of $F_c = 0\%$ with cyclic dilative response.

Figure 2.9a, b shows correlations $\sigma_d \sim \varepsilon$ and $\sigma_d \sim \sigma'_c$ for the original and fines-deprived sands, respectively. In the original sand with $F_c \approx 35\%$, strain tends to increase abruptly before reaching the pore-pressure builds up fully, while the sand with $F_c = 0\%$ behaves regularly as normal clean sands. Thus, the non-plastic fines

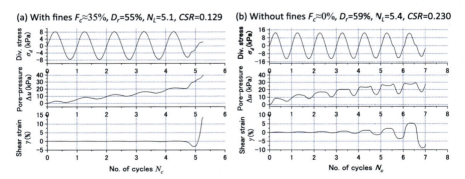

Fig. 2.8 Time histories of undrained cyclic triaxial tests on Sapporo sand: **a** Original sand with fines ($F_c = 35\%$), **b** sand deprived of fines ($F_c = 0\%$)

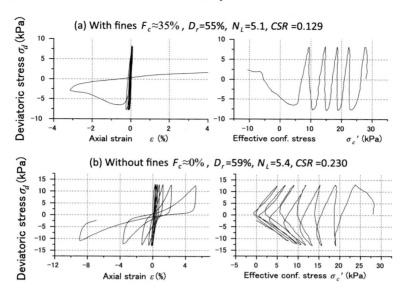

Fig. 2.9 Correlations $\sigma_d \sim \varepsilon$ and $\sigma_d \sim \sigma_c'$: **a** Original sand ($F_c = 35\%$), **b** Fines-deprived sand ($F_c = 0\%$)

seem to have distinctive effect on cyclic loading response in the Sapporo sand and presumably in the Kitami sand as well.

2.3.2 Monotonic Loading Response

In addition to the cyclic loading behavior, Fig. 2.10a, b shows undrained monotonic loading triaxial test results (in three-step effective confining stresses $\sigma_c' = 20, 40, 80$ kPa) of the same Sapporo sand of $F_c \approx 35\%$ and 0% for $D_r = 45\%$ and 70%, respectively, in terms of shear stress τ or pore-pressure Δu versus axial strain ε. The impact of F_c is remarkable not only for $D_r = 45\%$ but also for $D_r = 70\%$ as all the $\tau \sim \varepsilon$ curves of the original sand ($F_c \approx 35\%$) exhibit contractive behavior with post-peak strain softening in all σ_c'-values, while those of $F_c = 0\%$ behave in a dilative manner for lower σ_c'-values in particular.

The same trend can be observed in the effective stress paths of Fig. 2.11a, b, (derived from the same test results as in Fig. 2.10). For (a) $D_r = 45\%$, the original sand of $F_c \approx 35\%$ tends to be very contractive with the stress path undergoing monotonic strain-softening flow after taking peak values (when crossing a dashed CSR-line mentioned later) for all the σ_c'-values. In contrast, the same sand of $D_r = 45\%$ deprived of fines ($F_c = 0\%$) behaves quite differently with a clear turn (when crossing a PT line explained later) in the strain-softening path. For the high-density case (b) $D_r = 70\%$, the stress path behaves in a dilative manner from the start

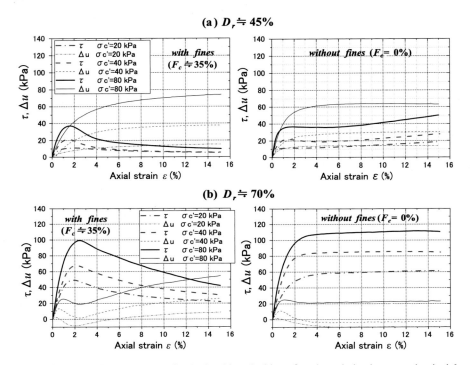

Fig. 2.10 Correlations τ (Δu) ~ ε for sands with and without fines in undrained monotonic triaxial tests: **a** $D_r = 45\%$, **b** $D_r = 70\%$

climbing up to a certain point and then take a sudden downturn in the original sand of $F_c \approx 35\%$ presumably due to collapsibility of soil skeletons caused by high fines content while no such downturn takes place in the sand of $F_c = 0\%$. Thus, it has been demonstrated that the inclusion of non-plastic fines in the Sapporo sand did considerably change the undrained shear behavior to be from dilative to contractive. The very similar change may well be estimated to have occurred in the Kitami sand, too.

2.4 Flow Failure Mechanism

2.4.1 Comparison with Torsional Shear Test

The effect of NP fines on the undrained cyclic loading failures under initial shear stress was discussed in the previous paper (Kokusho 2020) based on a series of torsional shear tests on reconstituted Futtsu beach sand near Tokyo. The tests were conducted on the isotropic effective confining stress of $\sigma'_c = 98$ kPa. The grain size

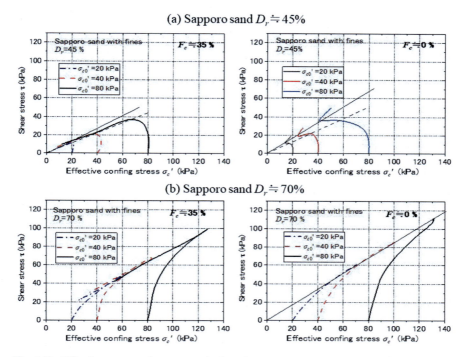

Fig. 2.11 Effective stress path on $\tau \sim \sigma_c'$ plane for sands with and without fines in undrained monotonic triaxial tests: **a** $D_r = 45\%$, **b** $D_r = 70\%$

curve of the Futtsu sand shown in Fig. 2.6b is very similar to the two case history sands in Fig. 2.6a and their fines are both non-plastic.

Figure 2.12 shows effective stress paths ($\tau \sim \sigma_c'$) on the left and stress-strain curves ($\tau \sim \gamma$) on the right obtained in monotonic shearing tests on the Futtsu sands with stepwise varying fines content $F_c = 0$–20%. Among the results of targeted

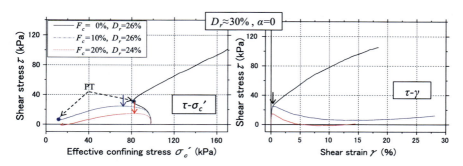

Fig. 2.12 Monotonic undrained shear test results on $\tau \sim \sigma_c'$ and $\tau \sim \gamma$ relationships of varying for $D_r \approx 30\%$, $\alpha = 0$ and $\sigma_c = 98$ kPa (Yamashita 2019; Yang et al. 2006)

relative density $D_r = 24 \sim 26\%$ in (a), the stress path for $F_c = 0\%$ shows dilative response, wherein σ'_c and τ both increase along a failure line after turning direction at a point PT corresponding to Phase Transformation (Ishihara et al. 1975). The associated $\tau \sim \gamma$ curve undergoes strain hardening after yielding at the arrow mark near the PT points. In contrast, the same sand with fines content $F_c = 10$ and 20% behaves contractively with strain-softening after taking peak stress shown with the arrows in the graph. For $F_c = 10\%$, the stress path goes down to a point corresponding to PT and turn up slightly, while the attendant $\tau \sim \gamma$ curve shows strain-softening behavior of limited flow down to a temporary minimum called quasi-steady state strength (Ishihara 1993) and recovers gradually thereafter. For $F_c = 20\%$, the stress path shows farther strain-softening after taking a lower peak, and moves toward the origin of zero effective stress, while the $\tau \sim \gamma$ curve approaches to zero residual strength.

The response of the Futtsu sand by the torsional simple shear tests in Fig. 2.12 may be compared to that of the Sapporo sand by the triaxial compression tests depicted in Fig. 2.11 despite the difference in test methods. The Sapporo sand of $F_c = 0\%$ shows dilative response with no strain-softening behavior for both $D_r \approx 70$ and 45%, that is similar to the Futtsu sand of $D_r \approx 30\%$, $F_c = 0\%$. With fines increasing to $F_c \approx 35\%$, the Sapporo sand changes to be very contractive with almost unlimited strain-softening for $D_r = 45\%$ in particular that is very similar to the Futtsu sand of $D_r \approx 30\%$, $F_c = 20\%$.

As a background of the above observations, let us go back to a fundamental shear mechanism associated with the void ratio (e) versus effective confining stress (σ'_c) State Diagram shown in Fig. 2.13. Though cyclic loading in level ground has been considered as a basis for liquefaction triggering mechanism since the pioneering research by Seed and Lee (1966), the sustained initial shear stress, responsible for serious liquefaction-induced consequences such as lateral spreading and flow sliding,

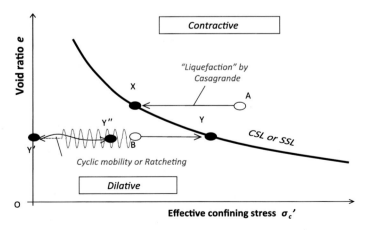

Fig. 2.13 $\sigma'_c \sim e$ state diagram and SSL (Steady State Line) or CSL (Critical State Line) dividing into contractive and dilative states

is absent there. Casagrande (1971) provided a different failure mechanism focusing on the role of the sustained initial shear stress in nearby slopes and superstructures. The same author, followed by Castro (1975), utilized the concept of the Steady State Line (SSL) or Critical State Line (CSL) on the State Diagram to interpret the failure mechanism. A soil element starting from point **A** on the contractive side of SSL tends to move leftward with decreasing effective confining stress σ'_c or decreasing stability, and eventually reaches steady-state flow at constant volume at **X** on the SSL as depicted in Fig. 2.13. If the soil is at **B** on the dilative side of SSL and monotonically loaded in the undrained condition, the point moves rightward to **Y** on SSL with increasing effective confining stress σ'_c where no destabilization occurs. If the same sand is loaded cyclically, the point moves from **B** to the left due to negative dilatancy during cyclic loading even in the dilative zone, different from positive dilatancy during monotonic loading, and eventually reaches zero-effective stress at **Y′** under the zero-initial shear stress condition in level ground as Seed and Lee (Casagrande 1971) demonstrated. However, if initial shear stress is considered, subsequent undrained monotonic loading translates the point rightward to **Y″** and the resistance of the sand revives again, restricting large flow-type deformation driven by the initial shear stress.

In normal liquefaction problems for depths of around 10 m or shallower, the effective confining stress is $\sigma'_c = 98$ kPa at most, and the relative density of loose sand deposits would be around $D_r = 30$–40% in the loosest case. In view of the test results in Fig. 2.12, clean sand ($F_c = 0$) is normally on the dilative side in ground depths for normal liquefaction evaluations, and the flow-type failure (even the limited flow-type) seems to be difficult to occur. A series of monotonic loading triaxial tests of clean Toyoura sand also support the same trend (Ishihara 1993).

However, the presence of NP-fines mixed with clean sands shifts the volume change behavior significantly. The effect of NP fines on the volume change and SSL was studied by Yang et al. (2006), Papadopoulou and Tika (2008), and Rahman and Baki (2011). They commonly demonstrated a considerable effect of fines, wherein the SSL tends to move down-leftward on the stress diagram, namely the contractive zone tends to expand toward smaller e and lower σ'_c, with increasing F_c up to around 30%.

2.4.2 Effect of Fines on Flow Failure During Cyclic Loading

Thus, it may well be judged that the increase of fines changed the Sapporo sand from dilative to contractive in the same way as the Futtsu sand with increasing F_c as shown above. Next, cyclic and monotonic loading test results are compared in the torsional simple shear tests on the Futtsu sand to know the effect of pertinent parameters on the cyclically induced flow failure under varying initial shear stress.

In Fig. 2.14a, c, the test results are shown in terms of $\tau \sim \sigma'_c$ (top) and the $\tau \sim \gamma$ (bottom) for $D_r \approx 30\%$, $F_c = 20\%$ and initial shear stress ratio varying stepwise as $\alpha = 0$, 0.035, 0.075, respectively. One monotonic and two cyclic loading tests

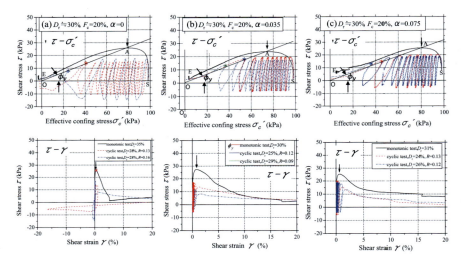

Fig. 2.14 Torsional simple shear test results of Futtsu sand in terms of $\tau \sim \sigma'_c$ (top) and the $\tau \sim \gamma$ (bottom) for $D_r \approx 30\%$, $F_c = 20\%$ and $\alpha = 0, 0.035, 0.075$

with different τ_d are chosen individually, which belong to the flow-type failure under initial shear stress except (a) $\alpha = 0$. The $\tau \sim \sigma'_c$ curve for the monotonic test in each graph indicates steady decrease of c all the way from the start (S) to the end (E), and the peak of stress τ at the point A is marked with an arrow. The straight line OA drawn from the origin O with the angle ϕ_y represents the CSR-line which was defined by Vaid and Chern (1985) as the initiating line of instable flow failure. The associated $\tau \sim \gamma$ curve takes clear stress peaks corresponding to the point A which are followed by strain-softening behavior leading to flow failure with monotonically declining shear strength.

In the cyclic loading tests, the $\tau \sim \sigma'_c$ curves in the top diagram undergo gradual effective stress decrease (or pore-pressure buildup) cyclically to certain points marked with * where sudden strain-softening sets off leading to unlimited flow failure thereafter to the point E. It should be pointed out in all the charts that the symbols * are located on or nearby the CSR-line AO defined by the corresponding monotonic tests.

This indicates that the CSR line represents the trigger of flow failure so that whenever the effective stress path comes across the line, strain-softening flow failure sets off irrespective of loading paths as long as the initial shear stress is working there. This observation serves as an experimental evidence in the simple shear stress condition of what (Vaid and Chern 1985) demonstrated in the triaxial stress state. In the bottom of Fig. 2.14, it is observed that corresponding strain exerted during cyclic loading is minor in magnitude and the major strain is attributed to the flow failure by strain-softening. Among the cases in Fig. 2.14 on the contractive side of SSL, the case (a) of $\alpha = 0$ undergoes non-biased, non-flow failure in the absence of initial shear stress. Also note that the flow failure of very gentle slopes with low α

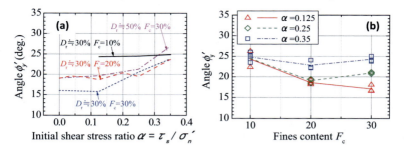

Fig. 2.15 Angle of CSR-line (Yang et al. 2006): **a** plotted versus initial shear stress ratio α, **b** plotted versus fines content F_c

values as 0.035 and 0.075 is triggered in the stress-reversal condition wherein the pore-pressure tends to build up faster.

In the above discussions, the initiation of brittle flow failure of saturated sand under initial shear stress is obviously governed by the yield condition named CSR-line on the $\tau \sim \sigma_c'$ diagram. In Fig. 2.15a, the angle ϕ_y determined from the torsional shear test results for sands of $D_r \approx 30, 50\%$ and $F_c = 10, 20$ and 30% are plotted versus the initial shear stress ratio α. The ϕ_y-value tends to be essentially constant against α increasing from zero to a certain limit, thereafter followed individually by an ascending trend. Figure 2.15b depicts the variations of ϕ_y against F_c for $\alpha = 0.125, 0.25$ and 0.35 for the same density $D_r \approx 30\%$, wherein the averages of two to three ϕ_y-values for identical F_c-values are connected with straight lines. The ϕ_y-value tends to decrease remarkably with increasing F_c for smaller α in particular, while F_c makes little difference in ϕ_y under the high value of $\alpha = 0.35$. This indicates that flow-type failure tends to be triggered more easily with increasing F_c in gentler slopes (the stress reversal is more likely to occur) than steeper slopes.

The observations above on the flow failure mechanism of the Futtsu sand mixed with NP fines may well be valid to the in situ sands of Sapporo and Kitami considering the similarity of their physical properties. Thus, the underground flow behavior in long-distance observed during recent two earthquakes in Hokkaido may essentially be accounted for by high contractility of in situ sands of high F_c undergoing unlimited flow in very gentle slopes.

2.4.3 Possible Scenario for Hokkaido Flow Failure Cases

Thus, a possible scenario of liquefaction-induced underground flows and associated ground depressions observed twice in Hokkaido Japan may be delineated as follows;

a. In a gently inclined fill, very loose sand containing large amount of NP fines under shallow water tables was cyclically sheared during earthquakes. When the stress path reached the CSR line, the sand started strain-softening behavior under the influence of initial shear stress.

b. Due to unlimited flowability of highly contractive sand leading to zero shear resistance, the sand mass fluidized even under the small slope gradient of 3% in the downslope direction, and thereby induced suction in its upper boundary compressing the unsaturated surface layer by atmospheric pressure. This may be able to explain why neither ground fissures nor sand boils were observed in/around the depressions unlike normal liquefaction cases.

c. The liquefied sand mass could keep flowing underground, though with low speed because of small driving force in gentle slopes, and ejected from a weak surface downslope in the margin of landfill. Consequently, non-liquefied surface layer was depressed to fill the underground cavity with little lateral movement.

d. This failure mechanism was made possible mainly because the liquefied sand was very flowable due to the large content of NP fines as demonstrated in the series of laboratory tests presented here, and possibly due to crushability of volcanic porous pumice particles. Water drainage systems that used to be at the bottom of the fill may have made some minor contribution to the flow-out in the case of Sapporo, while no such drainage pipes were available in the Kitami case.

e. The above-mentioned suction seems to cancel pore-pressure buildup, recover effective stress and thereby interrupt the sand to flow at least in the upper boundary of the liquefied sand. Nevertheless, the major portion of liquefied sand may have been able to flow presumably because overwhelming contractility tends to sustain 100% pore-pressure buildup inside the liquefied sand mass while its upper boundary was in suction at the same time, though more investigations are certainly needed to substantiate this mechanism.

2.5 Summary

Two case histories of strange liquefaction-induced flow failures in gently-inclined manmade fill during two recent earthquakes in Hokkaido, Japan were first outlined on their common features in failure modes, site and geotechnical conditions.

(1) They are characterized as gently inclined ($\approx$3%) landfills of low density with associated SPT N-value as low as unity in the extreme and of shallow groundwater.

(2) The failures accompanied neither ground fissures nor sand boiling but considerable depressions of exceeding 3 m because the underground liquefied sand fluidized long-distance downslope and erupted from selected point/points collectively.

(3) The two sands shared very similar physical properties; almost identical grain size curves, a lot of included NP fines exceeding $F_c = 30\%$, and extraordinarily low soil particle density reflecting high content of volcanic pumice.

Undrained triaxial tests were then conducted on the in situ sand sampled from the Sapporo site, and the results were compared with those by torsional simple shear tests on Futtsu sand of similar physical properties, wherein the effect of fines was focused.

(4) Cyclic loading liquefaction tests revealed that liquefaction resistance of the original Sapporo sand ($F_c = 35\%$) is extraordinarily low compared to the same sand deprived of fines ($F_c = 0\%$).

(5) Monotonic loading tests demonstrated that the Sapporo sand did considerably change the undrained shear behavior from dilative to contractive if compared between the original ($F_c = 35\%$) and that deprived of fines ($F_c = 0\%$). Similar change may well occur in the Kitami sand, too.

(6) In a series of torsional simple shear tests on Futtsu sand, the increase of NP fines tended to shift loose sand of $D_r \approx 30\%$ from being dilative to contractive, and the failure modes under initial shear stress tend to change correspondingly from non-flow cyclic failure to flow-type failure.

(7) The series of tests on contractive sands demonstrated that the flow failure sets off when the effective stress path comes across the CSR-line starting from the origin with angle ϕ_y on the $\tau \sim \sigma_c'$ diagram, which is uniquely determined for both monotonic and cyclic loading.

(8) The ϕ_y-value was found to decrease with increasing F_c particularly under small initial shear stress, indicating easier triggering of flow failure of high F_c-sands in gentle slopes in the stress-reversal condition.

A possible scenario of the two case history failures was constructed in the light of the laboratory tests as follows.

(9) Judging from the laboratory test results, considerable flowability of pumiceous fill sands are firstly attributable to the long-distance underground flow and attendant ground depression.

(10) NP fines contained in large amount made the loose artificially filled sands highly contractive and flowable on the contractive side of SSL under the effect of initial shear stress in gentle slopes $\approx 3\%$.

(11) Long-distance underground flow caused suction and eventually the ground depression compressed by atmospheric pressure without fissures, sand boils and lateral surface displacements.

(12) More detailed investigations by model tests and numerical analyses are needed to quantitatively verify this scenario and associated mechanisms.

Acknowledgements Ex-graduate students of Chuo University, Tokyo, Japan, Fumiki Ito, Takuya Kusaka and Ryotaro Arai who had conducted a series of laboratory tests for the last decade and generated the valuable database incorporated in this paper are acknowledged for their great contribution. Sapporo City Office, Professor emeritus Susumu Yasuda, Tokyo Denki University, and Professor Satoshi Yamashita, Kitami Institute of Technology are gratefully acknowledged for valuable information on the case histories.

References

Casagrande A (1971) On liquefaction phenomena. Geotechnique (London, England) XXI(3):197–202

Castro G (1975) Liquefaction and cyclic mobility of saturated sands. J Geotech Eng Div (ASCE) 101(GT6):551–569

Ishihara K, Tatsuoka F, Yasuda S (1975) Undrained deformation and liquefaction of sand under cyclic stresses. Soils Found 15(1):29–44

Ishihara K (1993) Liquefaction and flow failure during earthquakes, 33rd Rankine Lecture. Geotechnique 43(3):351–415

Ito Y, Yamashita S, Suzuki T, Hirata H (2005) Gentle-slope movements induced by the 2003 Tokachi-Oki Earthquake in the Kyowa area of Tanno Town Hokkaido Japan. Landslides-J Jpn Landslide Soc 42(2):103–111

Kokusho T (2020) Earthquake-induced flow liquefaction in fines-containing sands under initial shear stress by lab tests and its implication in case histories. Soil Dyn Earthq Eng (Elsevier) 130

NIED (2018) National Research Institute for Earth Science and Disaster Resilience, Tsukuba, Japan

Papadopoulou A, Tika T (2008) The effect of fines on critical state and liquefaction resistance characteristics of non-plastic silty sands. Soils Found Jpn Geotech Soc 48(5):713–725

Rahman MM, Lo SR, Baki MAL (2011) Equivalent granular state parameters and undrained behavior of sand-fines mixtures. Acta Geotech 6:183194

Sapporo City Office (2018) Publications for Forum to Residents on 2018 Earthquake Damage

Seed HB, Lee KL (1966) Liquefaction of saturated sands during cyclic loading. J SMFD (ASCE) 92(6):105–134

Tsukamoto Y, Ishihara K, Kokusho T, Hara T, Tsutsumi Y (2009) Fluidization and subsidence of gently sloped farming fields reclaimed with volcanic soils during 2003 Tokachi-oki earthquake in Japan. In: Geotechnical case history volume, Balkema, pp 109–118

Vaid YP, Chern JC (1985) Cyclic and monotonic undrained response of saturated sands, advances in the art of testing soils under cyclic conditions. In: Proceedings of the ASCE convention, Detroit, Mich., pp 120–147

Yamashita S, Ito Y, Hori T, Suzuki T, Murata Y (2005) Geotechnical properties of liquefied volcanic soil ground by 2003 Tokachi-Oki Earthquake, Published with Open Access under the Creative Commons BY-NC Licence by IOS Press

Yamashita S (2019) Personal communication

Yang S, Lacasse S, Sandven R (2006) Determination of the transitional fines content of mixtures of sand and non-plastic fines. Geotech Test J (ASTM) 29(2):102–107

Yasuda S (2018) Personal e-mail communications

Chapter 3
A Regional-Scale Analysis Based on a Combined Method for Rainfall-Induced Landslides and Debris Flows

Sangseom Jeong and Moonhyun Hong

3.1 Introduction

Large-mass and high-velocity debris flows can be fatal to human society and infrastructures. Solving these dynamic problems with extremely large deformations over a short event time is slightly different from traditional problems in geotechnical engineering. It is highly difficult to reproduce debris flows in the field, so lab- or large-scale experiments or numerical simulations are usually carried out to better understand the mechanisms and assess the risks (Chen and Lee 2000).

When numerically simulating debris flows, many numerical models are usually based on the Savage-Hutter theory (Savage and Hutter 1989). These flow models have to consider constant parameters such as the lateral earth pressure coefficient and the friction angle, which are usually acquired by independent experiments. However, these models are insufficient for reflecting the genuine behaviors of the debris mixture. Previous studies have proposed effective stress-dependent frictional resistance (Iverson 1997), a μ-parameterization model (Pouliquen and Forterre 2001), a thermo-pore-mechanical model (Vardoulakis 2000), and the velocity-dependent friction law (Liu et al. 2016).

Some studies have attempted to use a rheological model for non-Newtonian fluids or turbulent flows in shallow water equations to represent debris mixture behavior (Hong et al. 2020; Laigle and Coussot 1997). The rheological model is more flexible in representing the velocity-dependent resistance than the Coulomb friction model. Additionally, previous studies have reported that Coulomb frictional resistance from constant bed friction could be insufficient to dampen debris velocity (Hutter and Greve 1993). However, the viscous resistance in the rheological model could be much lower than the Coulomb frictional resistance when the debris flow has some thickness

---

S. Jeong (✉) · M. Hong
Yonsei University, Seoul 03722, Republic of Korea
e-mail: soj9081@yonsei.ac.kr

© The Author(s), under exclusive license to Springer Nature Singapore Pte Ltd. 2023
H. Hazarika et al. (eds.), *Sustainable Geo-Technologies for Climate Change Adaptation*, Springer Transactions in Civil and Environmental Engineering, https://doi.org/10.1007/978-981-19-4074-3_3

(Iverson 2003), and the estimation of rheological properties in debris mixtures has been less studied than in the Coulomb friction model.

Another problem facing rheological models of debris flows is that rheological properties are actually a function of the solid phase (Kaitna et al. 2007). This means that the solid volume fraction should be tracked to ultimately obtain more realistic and accurate results from a simulation considering the rheological model. Additionally, the mixture density cannot be constant when multiple debris flows with different densities merge at a confluence. However, most previous numerical studies have used single-phase models that assume a constant mixture density (Chen and Lee 2000). A few studies have recently suggested two-phase models for debris flows that contain continuity and momentum equations for both the solid and fluid phases (Pudasaini 2012). These two-phase models have a high potential to describe debris flow behaviors more realistically and overcome the limitations of current single-phase models. However, a theoretical basis with experimental evidence on the fluid-solid interactions employed in these two-phase models is still insufficient, and the models require greater computational effort and more input parameters than single-phase models.

To solve the shallow water equations for debris flows, many numerical studies have used the smoothed particle hydrodynamics (SPH) method and the finite volume method (FVM). SPH is a meshless and full Lagrangian-type approach that solves the individual dynamics of fictitious fluid particles by using Newton's second law. The fluid-fluid and fluid-solid interactions are applied to the particles by volume-averaging kernel functions that are macroscopically recovered by shallow water equations. However, SPH requires a sufficient number of particles to obtain an accurate solution, and the computational cost rapidly increases as the number of particles increases. Unlike SPH, the FVM is a mesh-based and Eulerian-type approach based on a divergence theorem that is specialized for computational fluid dynamics. Although SPH is flexible and can be recovered by shallow water equations, the numerical solutions obtained by SPH weakly and asymptotically satisfy the shallow water equations. In contrast, the FVM for debris flows requires sophisticated numerical treatments such as flux difference splitting schemes and can more strictly and accurately produce discontinuous solutions for shallow water equations.

This paper presents a simplified depth-averaged debris flow model for tracking density evolution. The developed model uses Hershel-Buckley rheology in internal and basal frictions and considers complex terrains and entrainments. In particular, the interaction between solid-fluid phases in the mixture is ignored. A finite volume formulation of the proposed model is presented with relevant numerical schemes to obtain stable and accurate solutions.

3.2 Methodology

The governing equations derived by depth-averaging the Navier-Stokes equations are used to simulate debris flows as fluid. The continuity equations for solid phase and debris defined as the mixture, and the momentum equations of the mixture phase for the *x*- and *y*-axis are given as

$$\frac{\partial h\rho}{\partial t} + \frac{\partial}{\partial x}(h\rho v_x) + \frac{\partial}{\partial y}(h\rho v_y) = -\rho_b \frac{\partial z_b}{\partial t} \quad (3.1)$$

$$\frac{\partial hc_s}{\partial t} + \frac{\partial}{\partial x}(hc_s v_{s,x}) + \frac{\partial}{\partial y}(hc_s v_{s,y}) = -c_{bs} \frac{\partial z_b}{\partial t} \quad (3.2)$$

$$\frac{\partial h\rho v_x}{\partial t} + \frac{\partial h\rho \alpha_m v_x v_x}{\partial x} + \frac{\partial h\rho \alpha_m v_y v_x}{\partial y} = +\frac{\partial h\overline{\tau}_{xx}}{\partial x}$$
$$+ \frac{\partial h\overline{\tau}_{yx}}{\partial y} - \frac{\partial h\overline{p}}{\partial x} - p_b \frac{\partial z_b}{\partial x} + \omega \tau_b t_{b,x} \quad (3.3)$$

$$\frac{\partial h\rho v_y}{\partial t} + \frac{\partial h\rho \alpha_m v_x v_y}{\partial x} + \frac{\partial h\rho \alpha_m v_y v_y}{\partial y} = +\frac{\partial h\overline{\tau}_{xy}}{\partial x}$$
$$+ \frac{\partial h\overline{\tau}_{yy}}{\partial y} - \frac{\partial h\overline{p}}{\partial y} - p_b \frac{\partial z_b}{\partial y} + \omega \tau_b t_{b,y} \quad (3.4)$$

where h is the debris height, ρ is the mixture density, and c_s is the solid volume fraction. v_k and $v_{k,s}$ are the depth-averaged velocity for each axis of the mixture and solid phases. α_m is a momentum correction factor and $\overline{\tau}_{ij}$ is the depth-averaged shear stress.

ω is a ratio of the basal surface area, and it can be written as

$$\omega = \sqrt{\left(\frac{\partial z_b}{\partial x}\right)^2 + \left(\frac{\partial z_b}{\partial y}\right)^2 + 1} \quad (3.5)$$

We assumed that the linear distribution of the flow velocity (parallel to basal surface) and the pressure (z-direction), then the bulk pressure p and the gravitational acceleration $\hat{g}$ are given as (Xia et al. 2013)

$$p = \rho \hat{g}(z_b + h - z) \quad (3.6)$$

$$\hat{g} = \frac{1}{\omega^2}[g + \boldsymbol{v}^T \boldsymbol{H} \boldsymbol{v}] \quad (3.7)$$

in which

$$v = \begin{bmatrix} v_x \\ v_y \end{bmatrix} \text{ and } \boldsymbol{H} = \begin{bmatrix} \frac{\partial^2 z_b}{\partial x^2} & \frac{\partial^2 z_b}{\partial x \partial y} \\ \frac{\partial^2 z_b}{\partial x \partial y} & \frac{\partial^2 z_b}{\partial y^2} \end{bmatrix} \qquad (3.8)$$

where $\boldsymbol{H}$ is the Hessian matrix of the surface elevation.
The basal friction stress of the fluid can be given as

$$\tau_b = \mu \frac{2 - \alpha}{(1 - \alpha)} \frac{\sqrt{v_x^2 + v_y^2 + v_z^2}}{h} \qquad (3.9)$$

where μ is the debris viscosity.
This study uses the Herschel-Buckley model for the rheology of debris flow as

$$\mu = \min\left(\left[\tau_y + k_0 \gamma^n\right]\gamma^{-1}, \mu_{\max}\right) \qquad (3.10)$$

where τ_y is the yield stress, γ is the magnitude of the shear rate, k_0 is a consistency index, n is a flow index, and $\mu_{\max}$ is the maximum viscosity that prevents infinitely high viscosity when the shear rate approaches zero.

3.3 Modeling of Debris Flows in a Mountainous Area

3.3.1 Study Area Description

The study area is the catchment No. 30 in the mountains above Yu Tung Road, in the southeast of Tung Chung New Town in Lantau (Fig. 3.1). The top of the hill in the study area is sloping between 30° and 45° and has a locally steep rock exposure area. In the midstream, hill slopes are typically between 15° and 30°, and down to less than 15° in bottom of the stream. Several debris flows were occurred at around 9:00 a.m. on June 7, 2008, and the debris flow occurred in the catchment No. 30 is the largest debris flow in the mountainous area near the Yu Tung Road. The volume of the landslide source at the top of the watershed was measured to be 2,350 m^3 and the debris flowed into the adjacent drainage line. The maximum activity volume by entrainments is observed to increase to about 3,400 m^3, and the run-out distance was estimated to be about 600 m. The landslide area was located at the elevation of 202 m southwest of the drainage line under the rocky outcrop. Based on the post-landslide topographic surveys, the landslide source included a lot of gravels of about 2,350 m^3, with some silty of clayey sand and rocks. The slope failure area was about 32 m × 50 m, and the maximum thickness was reported to be about 3 m. The slope of the failure surface varied between 35° and 50°. All these descriptions were available from the GEO report No. 271 (2012) published by Geotechnical Engineering Office (GEO), Hong Kong.

3 A Regional-Scale Analysis Based on a Combined Method … 39

Fig. 3.1 The study area of Yu Tung Road historical landslides and debris flow case

3.3.2 Modeling and Input Parameters

The historical debris flow, 2008 Yu Tung Road debris flow, was simulated based on the provided topography and initial thickness of the landslide. Initial soil depth of the study area was assumed from 3 to 16 m based on the previous ground investigation of detailed study report of the study area (Kwan 2012). Internal friction angle of residual soil was determined as 30° based on the previous study conducted by Law et al. (2017). Law et al. (2017) performed a series of back analyses on the debris flow in Yu Tung Road, 2008 using 3d-DMM model, and compared the analytical results with field observations. Basal friction angle and drag coefficient were determined as 11° and 500 m/s^2 based on the back-analysis results of GEO report No. 271 (2012). In this study, a series of back analyses was also performed previously to determine the initial dynamic viscosity of debris flows, and the initial dynamic viscosity of 0.1 Pa s was adopted in the debris flow modeling. Parameters for the simulation were summarized in Table 3.1. Figure 3.2a shows the elevation of initial state of study area, and the initial volume was applied at the top of the watershed from the detailed investigation for the debris flow.

3.3.3 Results and Discussion

Debris profiles and changes of the elevation by entrainments and sediments at each representative times are shown in Fig. 3.2. As shown in Fig. 3.2b, f, soil erosions and entrainments were occurred over the path of the debris flow, and some debris

Table 3.1 The input data set and parameters for the Yu Tung road case

Contents		Data and value
Topographical data	Elevation (m)	Depends on modeling
	Soil depth (m)	Distributed from 3 to 16
Soil properties	Dry unit weight (kN/m^3)	18
	Saturated unit weight (kN/m^3)	20
	Cohesion (kPa)	0
	Inter friction angle (°)	30
Vegetation properties	Root cohesion (kPa)	0 (no vegetation assumed)
	Tree load (kN/m^2)	0 (no vegetation assumed)
	Interception loss by leaf (%)	0 (no vegetation assumed)
Fluid properties	Basal friction angle (°)	11
	Drag coefficient (m/s^2)	500
	Initial dynamic viscosity (Pa · s)	(back-analyzed)

materials were deposited on the drainage line. To analyze the erosion and deposit over the flow path of the debris flow, the time-varying volume of debris flow and the debris volume versus debris front position relation are also shown in Fig. 3.3 (Kwan 2012). The volume of the debris flow increased from the initial 2,350 m^3 to a maximum of 3,480 m^3, and gradually decreased as it deposited. The initial volume was fixed, and changes in the volume of debris flow are slightly different in the intermediate zone from about 200 to 350 m. Although there is some difference in intermediate zone, similar trends in the analytical results with observed volume and the maximum volume of debris flow are about to equal each other. Comparisons of time-varying front location and the front velocity with previous studies and field observed values are shown in Fig. 3.4. The analytical results by previous research (Law et al. 2017; Dai et al. 2017; Koo et al. 2017) are also compared with the results of this study and measured data reported in Geo report No. 271 (Kwan 2012). In comparison of the analytical results and measured data, the proposed method in this study slightly overestimated the time-varying front position of debris while that in the previous study also lightly underestimated, but the difference between the analytical results and measured data was very small (Fig. 3.4a). The front velocity of debris flows estimated in this study shown in Fig. 3.4b was also compared with

3 A Regional-Scale Analysis Based on a Combined Method ...

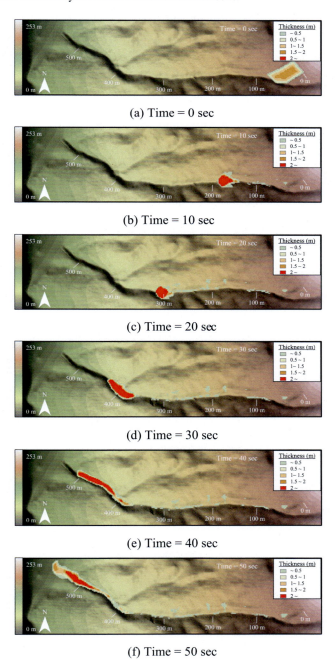

(a) Time = 0 sec

(b) Time = 10 sec

(c) Time = 20 sec

(d) Time = 30 sec

(e) Time = 40 sec

(f) Time = 50 sec

Fig. 3.2 Time-varying debris flow thickness of the study area

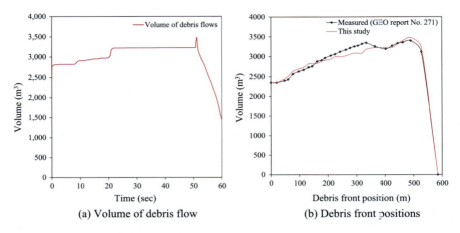

Fig. 3.3 Time-varying the volume of debris flow and debris front positions

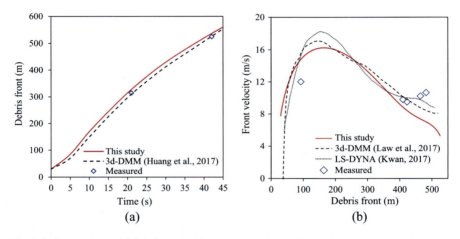

Fig. 3.4 Comparison of debris front positions and velocities with previous study

the results of the previous research and the measured data. It is shown that the front velocity of debris flows by this study has a similar trend with the results from previous study and measured data. These numerical results are greatly governed by the input parameters, which are very close to the measured values because the input parameters were determined by the back-analysis.

As a result of this watershed-scale historical debris flows case, it has been confirmed that the proposed method in this study can predict reasonably the mobility of debris flows (including the velocity and the front position) and the volume changes of debris reasonably in the actual land-scale debris flow prediction.

3.4 Conclusions

This work developed a simplified debris flow model with Hershel-Buckley rheology for tracking density evolution. A finite volume formulation of the debris flow model was also proposed for accurate and stable numerical simulations. Both the internal and basal frictions of the debris flow were considered in the model as well as the basal topology effect. A case of debris flows simulation was performed to validate the proposed method. One of the calibration cases is a field-scale experiment reported by a previous study, and it was used to compare the results of the proposed method with the previous study on the field-scale experiment. One of the real landslides and debris flow cases was simulated to compare the results by the proposed method with observations. By using the model developed in this study, it is possible to simulate not only reverse analysis after events but also the expandable debris flow induced by the input rainfall applied in engineering practice.

References

Chen H, Lee CF (2000) Numerical simulation of debris flows. Can Geotech J 37(1):146–160
Dai Z, Huang Y, Cheng H, Xu Q (2017) SPH model for fluid–structure interaction and its application to debris flow impact estimation. Landslides 14(3):917–928
Hong M, Jeong S, Kim J (2020) A combined method for modeling the triggering and propagation of debris flows. Landslides 17(4):805–824
Hutter K, Greve R (1993) Two-dimensional similarity solutions for finite-mass granular avalanches with Coulomb-and viscous-type frictional resistance. J Glaciol 39(132):357–372
Iverson RM (1997) The physics of debris flows. Rev Geophys 35(3):245–296
Iverson RM (2003) The debris-flow rheology myth. Debris-Flow Hazards Mitig: Mech Predict Assess 1:303–314
Kaitna R, Rickenmann D, Schatzmann M (2007) Experimental study on rheologic behaviour of debris flow material. Acta Geotech 2(2):71–85
Koo RCH, Kwan JS, Ng CWW, Lam C, Choi CE, Song D, Pun WK (2017) Velocity attenuation of debris flows and a new momentum-based load model for rigid barriers. Landslides 14(2):617–629
Kwan JSH (2012) Supplementary technical guidance on design of rigid debris-resisting barriers. Geotech Eng Office, HKSAR. GEO Report (270)
Laigle D, Coussot P (1997) Numerical modeling of mudflows. J Hydraul Eng 123(7):617–623
Law RP, Kwan JS, Ko FW, Sun HW (2017) Three-dimensional debris mobility modelling coupling smoothed particle hydrodynamics and ArcGIS. In: Proceedings of the 19th international conference on soil mechanics and geotechnical engineering, Seoul, pp 3501–3504
Liu W, He S, Li X, Xu Q (2016) Two-dimensional landslide dynamic simulation based on a velocity-weakening friction law. Landslides 13(5):957–965
Pouliquen O, Forterre Y (2001) Friction law for dense granular flows: application to the motion of a mass down a rough inclined plane. arXiv preprint cond-mat/0108398
Pudasaini SP (2012) A general two-phase debris flow model. J Geophys Res: Earth Surf 117(F3)

Savage SB, Hutter K (1989) The motion of a finite mass of granular material down a rough incline. J Fluid Mech 199:177–215

Vardoulakis I (2000) Catastrophic landslides due to frictional heating of the failure plane. Mech Cohesive-Frict Mater: Int J Exp Modell Comput Mater Struct 5(6):443–467

Xia X, Liang Q, Pastor M, Zou W, Zhuang YF (2013) Balancing the source terms in a SPH model for solving the shallow water equations. Adv Water Resour 59:25–38

Chapter 4
Views on Recent Rainfall-Induced Slope Disasters and Floods

Ikuo Towhata

4.1 Introduction

Many sizeable rainfall disasters have happened in Japan during the recent decade. In 2013, Izu Oshima Island (Fig. 4.1) was affected by the Typhoon No. 26 "Wipha" that brought 824 mm of rainfall in one night and triggered substantial slope failure in midnight (Towhata et al. 2021). This disaster (Fig. 4.2) notably occurred in a volcanic slope that had been stable since its formation by lava flow in 1338. Kyushu Island of Japan where Fukuoka City is located has suffered heavy rain disasters in 2009, 2012, 2016, 2017, 2019, and 2020 in recent years. These events give us an impression that the risk of rainfall disasters is getting higher nowadays and that this situation is possibly related with the global warming and climate change.

The increasing risk has been pointed out elsewhere. For example, JMA (Japan Meteorological Agency) interprets rainfall observation records (by AMeDAS network) to show that the occurrence of hourly rainfall exceeding 50 mm/h is increasing (Fig. 4.3). Moreover, the JMA's AMeDAS records of rainfall indicate that the daily rainfall exceeding 400 mm/day is more frequent in the twenty-first century (Fig. 4.4). Accordingly, the top 20 of JMA daily rainfall has been increasing in the recent years (Fig. 4.5). It is further interesting in Fig. 4.5 that there was no remarkable rainfall event in the middle of 2010s in spite of the aforementioned increasing trend in Fig. 4.3. In line with this, Grinsted et al. (2019) provided the area of total destruction by hurricanes in USA since 1900 (Fig. 4.6). While the range of variation is profound, the regression of data suggests that the size of the destroyed area is increasing with years.

Fukuoka Prefecture is vulnerable to rainfall disaster. Figure 4.7 by the Ministry of Land, Infrastructure, Transport and Tourism (abbreviated as MLIT hereinafter)

I. Towhata (✉)
Department of Civil Engineering, Kanto Gakuin University, Yokohama, Japan
e-mail: towhata.ikuo.ikuo@gmail.com

Fig. 4.1 Location of places that are referred to in this paper

Fig. 4.2 Rainfall-induced slope failure in Izu Oshima Island, Japan, in 2013

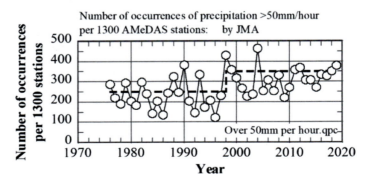

Fig. 4.3 Number of occurrences of hourly rainfall greater than 50 mm/h observed by JMA AMeDAS network (drawn by the author after JMA data)

4 Views on Recent Rainfall-Induced Slope Disasters and Floods

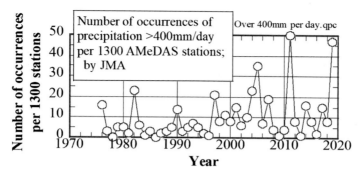

Fig. 4.4 Number of occurrences of daily rainfall greater than 400 mm/day (drawn by the author using data by JMA AMeDAS network)

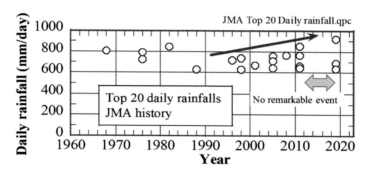

Fig. 4.5 Top 20 of daily rainfall records (data by JMA)

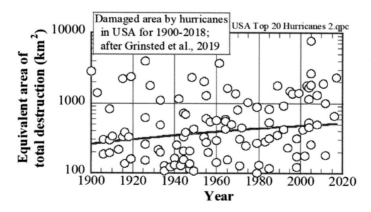

Fig. 4.6 Size of totally destroyed area during hurricanes in USA since 1900 (drawn after Grinsted et al. 2019)

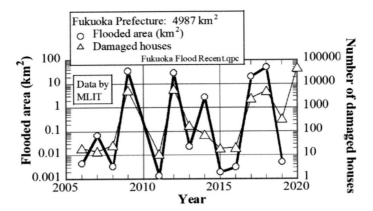

Fig. 4.7 History of flood disasters in Fukuoka Prefecture (drawn after MLIT flood Statistics)

exhibits that notable disasters have happened quite often in the past 10 years and, in particular, after 2017. Figure 4.8 shows the history of the maximum daily rainfall of the year at three cities, which are Fukuoka and Asakura in Fukuoka Prefecture together with Hita in Oita Prefecture; see Fig. 4.1 for their locations. It appears that there is an increasing trend in the twenty-first century with a significantly increasing trend in Asakura after 2015.

Many ongoing studies on rainfall-induced disasters focus on the effects of global climate change. Although this viewpoint is important, the author feels that the effects of natural action (rainfall intensity) and mitigation by human efforts (performance of dams, levees, slope reinforcement, etc.) should be discussed separately because, as is the convention in infrastructure design, the extent of safety depends on the magnitudes of action and mitigation. In other words, disaster will not be aggravated if mitigation is improved faster than the change of global climate.

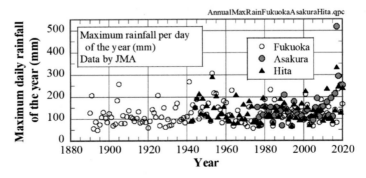

Fig. 4.8 History of the maximum daily rainfall of the year in Fukuoka, Asakura, and Hita (data by JMA)

4.2 Notes on Global Climate Change

The effect of global climate change is one of the hot points of discussion today. Figure 4.9 presents one of the temperature records that indicates the continuous warming throughout the twentieth century and supports the idea of global warming. From the viewpoint of disaster mitigation, the global warming has been discussed in the context of melting ice, sea level rise, and the likelihood of heavy rainfall. Furthermore, some people notably consider the global climate change as a serious threat to human community. In this regard, this chapter pays attention to the current extent of threat induced by climate change.

4.2.1 Intensity of Rainfall

Figure 4.3 referred to the JMA's AMeDAS data that indicates the increased likelihood of heavier 1 h rainfall. Because the strongest 1 h rainfall is not necessarily strong enough to trigger profound rainfall disaster, more elongated rainfall (24 h rainfall exceeding 400 mm/day) was illustrated in Fig. 4.4. It was therein shown that once per several years in the twenty-first century (2005, 2011, 2019) the number of heavy rainfalls is substantial. It is important that the annual precipitation has not increased so significantly contrary to the short-term concentrated rainfall (Fig. 4.10).

4.2.2 Temperature of Sea Water

Warmer sea water promotes more water evaporation and increases humidity in air and precipitation. In the typhoon season, the higher sea water brings stronger storms as well. In this perspective, Fig. 4.11 examines the temperature of sea water to the

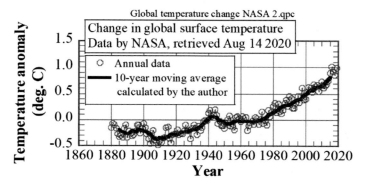

Fig. 4.9 Long-term rising of temperature (data by NASA)

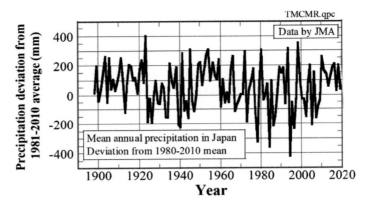

Fig. 4.10 Mean annual precipitation in Japan (drawn by using JMA data)

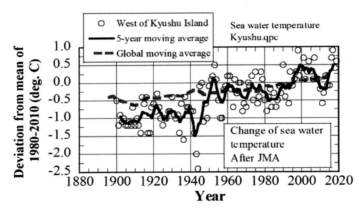

Fig. 4.11 Warming of sea water to the west of Kyushu Island (drawn by using JMA data)

west of Kyushu Island. The 5-year moving average shows that the sea water in this area is getting warmer in line with the global moving average. Thus, the risk of disaster in Kyushu is increasing.

4.2.3 Historical Information on Temperature in Winter

The weakness of the discussion on long-term warming as shown in Fig. 4.9 is the lack of old data. Instrumental record of temperature started in nineteenth century and the current warming trend might be a short-term fluctuation. Note, however, that the fluctuation of climate and precipitation are serious as well in disaster mitigation and the author does not mean that only the long-term climate change increases the

(a) Entire view of ice rampart   (b) Detail

Fig. 4.12 Ice rampart or *Omiwatari* in frozen lake Suwa; pictures were taken on February 2, 2018

rainfall hazard. To date, the studies on the climate change in the past several centuries relied on interpretation of historical information that is proxy of temperature.

The lake Suwa (Fig. 4.1) is located at 759 m above sea level in the mountainous region of Japan. In winter, the lake used to get frozen completely and the ice developed cracks and liftings across the lake, which is called ice rampart (ice ridge or *Omiwatari*); see Fig. 4.12. Local people consider this mythical phenomenon as journey of god across the lake and conduct ceremony upon its occurrence. This is a long tradition of the local community, and there is a record on the day of ice rampart formation in 1398 and then since 1444 until today.

Fujiwara and Arakawa (1954), followed by Arakawa (1954), stated that the recorded date of ice rampart formation implies the temperature in the winter of the year; later formation of rampart means warmer winter. Accordingly, it was thought that the date of ice rampart may be an index of global warming over centuries (Arakawa 1954).

The process of rampart formation is not so simple as volume expansion of water upon freezing as may be imagined. Arakawa (1954) stated that rampart is formed when the air temperature is lower than $-10\,^{\circ}$C, the entire lake gets frozen, the low temperature continues for several days, and the ice surface is prone to extremely low temperature in the air. After Hobbs (1911) and Tanaka (1918), Sugimoto et al. (1981) stated that cracks open in ice under extremely low temperature, the cracks are filled with water that is then frozen but forms relatively week parts of ice. The ice expands laterally in daytime due to thermal expansion, thereinafter associated with breakage, and uplifting of the weak parts (rampart formation being similar to buckling). The shortcoming of the historical dates of ice rampart formation as an index of global climate change is that documents in different eras record different types of dates such as that of completion of lake freezing, formation of ice ramparts, report to government, and religious ceremony (heterogeneity of data according to Ishiguro et al. 2002). In case of 2018, the lake was completely frozen on January 27, followed by ice rampart formation on February 2 and the ceremony on February 5. Thus, the historical record is not so accurate to allow discussion of 1 week difference over age.

Fig. 4.13 Number of years of ice rampart formation in lake Suwa in 10-year period (data by Fujiwara and Arakawa 1954; Yoneyama 1988, and Wikipedia for more recent years)

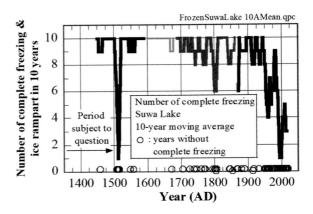

To avoid the above-mentioned problem, the author simply classified the record into the years with and without ice rampart. This viewpoint is not affected by the types of rampart dates. Figure 4.13 illustrates the moving average of the number of years in 10-year period in which lake was completely frozen or ice rampart was formed or any official report or ceremony took place. This figure further illustrates the years when the lake did not get frozen completely (warmer years). It is found that the number of freezing years decreased or non-freezing years increased in the twentieth century. The number of freezing years started to decrease after AD1800 as well. Although the warming trend herein appears evident in the twentieth century, further care is necessary of the possible effects of water contamination in the lake (depression of freezing point) and local heat-island effect as well as the infrastructure development along the lake shore.

More study on old climate deserves attention. Kajander (1993) reported the date of ice melting in the Tornio River at the border between Sweden and Finland. His idea was that the earlier melting day in May implies warmer climate. This data was plotted in Fig. 4.14 together with its 10-year moving average. It is reasonable to say that the date of melting started to become earlier probably near the end of the nineteenth century. This finding is consistent with Sharma et al. (2016). Another set of data comes from the weather monitoring in De Bilt, the Netherlands. Van den Dool et al. (1978) reconstructed the average winter temperature (December–February) since 1634. This data was combined with additional data for later years from "wunderground" to draw Fig. 4.15. It is again seen here that winter temperature started to rise near the end of the nineteenth century. Accordingly, the temperature of the world has been probably rising since the nineteenth century. The author, however, does not say anything with its relationship with the CO_2 emission. Note that the discussion on Figs. 4.13, 4.14 and 4.15 is concerned only with the winter temperature and has nothing to do with weather in summer. Furthermore, the author is not confident whether or not the warming trend will continue in future and affect the severity of rainfall-induced damage.

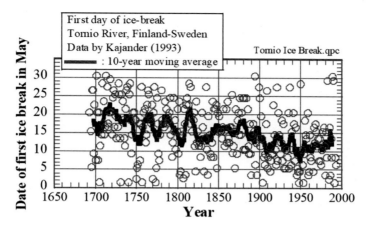

Fig. 4.14 Date of ice melting in the Tornio River (data by Kajander (1993))

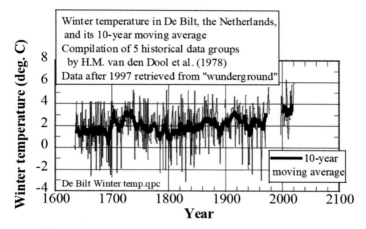

Fig. 4.15 Mean temperature in winter in De Bilt, the Netherlands (data by Van den Dool et al. (1978) with supplementary data from wunderground; https://www.wunderground.com/history/wee kly/nl/de-bilt/IDEBILT13/date/2019-11-12 retrieved on August 20, 2020)

4.2.4 Sea Level Rise

Rising sea level is another topic in global climate discussion and potentially promotes the risk of high sea water or surge. The sea level is caused to rise by melting ice as well as thermal expansion of water. Changes in wind and sea-current conditions are other agents to affect the sea water level. The question here is whether or not the sea level is rising and, if the answer is yes, how significant it is.

The Itsukushima Shrine is situated on shoreline near Hiroshima. Because of its unique location, this shrine is prone to sea wave action and inundation (Fig. 4.16).

Fig. 4.16 Itsukushima Shrine near Hiroshima located on shore line

Toyota (2011) stated that the frequency of inundation of this shrine has increased in the recent years. With regard to this situation, MLIT (2008) summarized the records of recent inundation to illustrate the increasing risk of this shrine (Fig. 4.17). To know whether or not this risk is related with the sea level rise, the author collected the record of annual sea level at the nearby Hiroshima Tide Station (upper part of Fig. 4.18). This figure evidently illustrates the rising trend after 2008. Care must be taken of this data, however, because the site of the observation is prone to long-term consolidation settlement. Therefore, the data was corrected by using the consolidation settlement of 0.512 cm/year proposed by Umeki (2003) who hypothesized that the recorded tide level change until 2003 was solely caused by consolidation. After subtraction of this consolidation component, the remaining value of tide level is plotted in the lower half of Fig. 4.18. It is herein shown that sea level rise is negligible throughout the studied period. Thus, it may be said that the inundation of the Itsukushima Shrine is caused by abnormal wave and wind actions that are becoming predominant in the recent times. This conclusion is consistent with that of Suenaga et al. (2003).

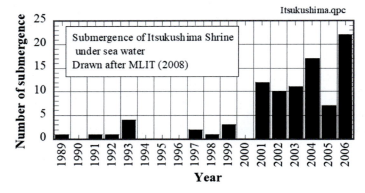

Fig. 4.17 Frequency of inundation of Itsukushima Shrine in the recent times (data by MLIT, 2008)

4 Views on Recent Rainfall-Induced Slope Disasters and Floods

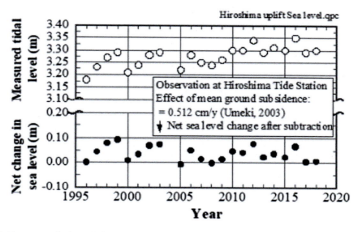

Fig. 4.18 Recent variation of tide level at Hiroshima Tide Station

Inundation of the sightseeing spots in Venice, Italy, attracts universal concern (Fig. 4.19). Being called *Acqua Alta* (high water), the floods in Venice are considered to be one of the symbols of global climate change. Figure 4.20 was drawn by using data provided by City of Venice (Cittá di Venezia). It is indicated herein that *Acqua Alta* occurs more frequently in the recent years. Groundwater pumping and consolidation settlement were the major threat in Venice. Therefore, pumping was banned in 1970s but sea level has been still rising. Carbognin et al. (2009) studied the

Fig. 4.19 Venice in normal tourist season in 2000

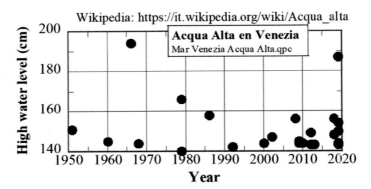

Fig. 4.20 Occurrence of high water and flood in Venice (data by Cittá di Venezia)

northern Adriatic Sea to show that the sea level in the region is continuously rising. They suggested that the rate of 0.12 cm/year of sea level rise at the stable station in Trieste (probably free of consolidation) is the effect of global climate change. This rise is superimposed by the strong wind and promotes the risk of *Acqua Alta* in Venice.

The delta of Bangladesh is prone to high wave upon cyclone landing. Few data is available on sea level rise in this region. The extremely high wave (surge) since late 1970s was studied by Lee (2013) to show that there is no systematic change until 2010, except events in 2005 and 2006. Furthermore, Fig. 4.21 indicates the recent history of wind speed during cyclones. There seems to be no increasing trend.

The data so far shown suggests that the threat of sea level rise is not serious at this moment except the high water level in Venice. Note, however, that the problem in Venice is not solely due to sea level rise. Its compound nature with consolidation

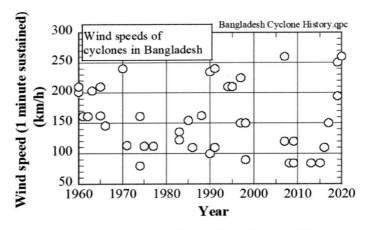

Fig. 4.21 Wind speed during cyclones in Bangladesh (data by Banglapedia)

settlement and wind conditions has to be recognized. In contrast, the scope toward future is not clear. If small island countries are prone to sea level rise, engineering has to propose inexpensive measures to protect the island communities, including relocation and immigration of the entire community. It deserves attention that many coral islands were formed by carbonate production and growth of coral reef, while the islands sank into the sea and/or sea level has risen more than one hundred meters after the last glacial period. The rate of coral reef growth was measured by Tanaya et al. (2018) in Okinawa, Japan. They found the rate of growth to range from 1.2 ± 0.4 to 3.2 ± 0.7 mm/year or 3.9 ± 1.5 mm/year, depending on methodology of observation. If this rate exceeds that of (sea level rise) + (ground subsidence), those islands are unlikely to sink into the sea. It is interesting that this rate of coral growth is comparable with the aforementioned rate of sea level rise of 0.12 cm/year at Trieste. Is it possible that coral grows together with the rising of sea level?

4.3 History of Rainfall-Induced Disasters in Japan from 1945 to 2020

Most discussion on the current increasing trend of rainfall-induced disasters has been made from the viewpoint of rainfall impact. In contrast, little attention has been paid to disaster mitigation infrastructures. The memory of recent heavy disasters gives us impression that more disasters are occurring in the recent years than in the past. However, the memory of recent events is always clearer than the memory of old events and our impression is biased by the recent memory. Furthermore, one should note that disaster is a consequence of the rainfall whose influence exceeds the resistance of mitigative structures. Because of these problems, the discussion in the past was not good enough. Accordingly, the author attempts here to examine the history of disasters in Japan since 1945 in order to understand whether or not the rainfall-induced disasters are increasing nowadays. Most data was collected from "Flood Statistics" published annually by MLIT and JMA website: http://www.jma.go.jp/jma/menu/menureport.html.

It was necessary to define a suitable index that accounts for the size of disasters. It was expected that the temporal change of the index under increasing trend of rainfall intensity demonstrates the long-term trend of risk of disasters. Among many candidate indices, the number of fatalities was ruled out because considerable efforts are going on today to let people evacuate from disaster-prone areas irrespective of the size of flood. Figure 4.22 illustrates the reduced fatality after tremendous efforts for the past 50 years. Nowadays, it is very possible that few people are killed in spite of large area of inundation. Then, search was made to find such indices that are easily available from both recent and past disasters. The damage in terms of monetary unit is popular but it is subject to inflation and not suitable. Finally, two indices were chosen which are the number of damaged houses and the area of inundation. While there are different kinds of house damage such as full collapse,

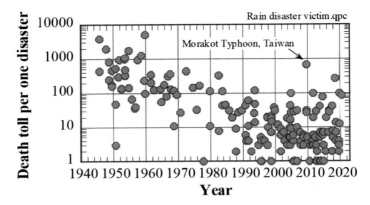

Fig. 4.22 Number of fatalities during recent heavy rain disasters

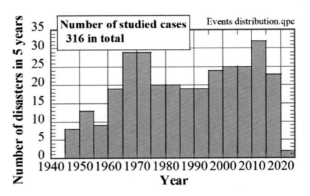

Fig. 4.23 Distribution of studied rainfall disasters over years

half collapse, and inundation above and below floor, the present study employs the summation of all types of damaged houses. JMA (Japan Meteorological Agency) announces the number of damaged houses after all major disasters since September 1945. Similar data is available in "*Flood Statistics*" that has been published by MLIT (Ministry of Land, Infrastructure, Transport and Tourism) since 1962. Because JMA and MLIT employ different practice of damage investigation, they publish different number of damaged houses. The other damage index is the area of inundation (square kilometers). JMA used to publish data but nowadays only MLIT does it. Those two indices are presented in what follows and then will be compared against the rainfall intensity data published by JMA.

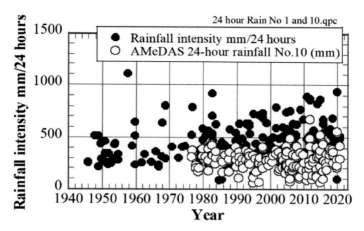

Fig. 4.24 Maximum daily or 24-h rainfalls during disasters (data by JMA)

4.3.1 Rainfall Intensity that Caused Disasters

Figure 4.23 shows the number of studied disasters per 5 years. In total, documents on 316 disasters were collected and studied. Figure 4.24 plots the maximum rainfalls per day starting at 0 AM or during 24 h starting at any time of the day. Note that what is plotted here is the maximum rainfall intensity during each disaster in order to shed light on what kind of rainfall triggered disasters. JMA used to report rainfall data observed only at its observatories. Since 1974, JMA has been operating 1300 remote stations over the nation under the name of AMeDAS so that more detailed rainfall distribution may be captured. The AMeDAS system is able to detect locally concentrated rainfall and let JMA publish top 10 (or 20) rainfalls during each disaster. Because the heaviest rainfall data may be too extreme, the use of No. 10 record (shown by hollow circles) in Fig. 4.24 may help us study reasonable rainfall intensity over disaster-hit region. Since 1985, the upper bound of both No.1 and No.10 rainfalls has been increasing. This increasing trend in the recent decades is recognized in the total rainfall during respective disasters (Fig. 4.25) as well. Thus, disasters have been triggered by stronger rainfalls.

4.3.2 Number of Damaged Houses During Rainfall Disasters

As stated before, the present study employs two types of damage indices that account for the size and severity of induced rainfall disasters. They are namely the number of damaged houses, consisting of destroyed houses and inundated houses, and the area of inundation. These data have been published by two institutions independently:

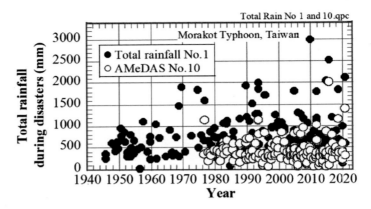

Fig. 4.25 Maximum total rainfalls during disasters (data by JMA)

JMA and MLIT. The present study addresses those two kinds of data and compare. JMA also measures and provides rainfall information more in detail than MLIT.

Figure 4.26 compares JMA and MLIT data. Although they are consistent with each other, there are cases where there is a significant difference and MLIT data is substantially greater than JMA data. One of the possible reasons for this difference is that the two institutions collect data through different channels and summarize them with different classifications, JMA with local administrative areas and MLIT with river basins. Another possible reason is that MLIT data is updated many times and finalized longer after the disasters than JMA. Furthermore, MLIT data is available only after 1962. At this moment, the author cannot decide which type of data is more relevant for the present study and, hence, employ both of them with due distinction.

Figure 4.27 illustrates the number of house damage reported by JMA. Until 1970s, there were many serious disasters in which more than 400,000 houses were affected in a single rainfall event (mostly within 24 h). This bad situation changed after 1980

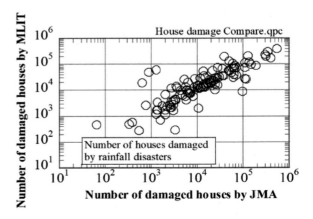

Fig. 4.26 Comparison of numbers of damaged houses reported by JMA and MLIT

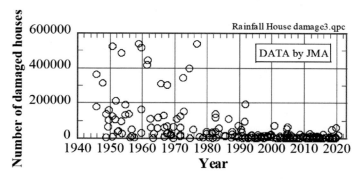

Fig. 4.27 Number of houses damaged by rainfall disasters (data by JMA)

and the upper bound of the number of damaged houses decreased. This situation might be consistent with the decrease of 24-h rainfall in 1980s (Fig. 4.24) or a consequence of long-term human efforts toward disaster mitigation. Another issue is that the number of damaged houses hit the minimal in the middle 2010s and, although being not obvious, the number is increasing after 2016. These points are more clearly shown in Fig. 4.28 by MLIT. Particularly, the number of damaged houses attained the minimal in the middle 2010s, followed by an increasing trend. At this moment, the author is not confident whether or not the recent increasing trend is meaningful; it might be data scattering. Figure 4.29 combines JMA and MLIT data together and demonstrates the situation only after 1980. The minimum level in the middle 2010s and increase in more recent years are visible here. Then the question is whether this trend is meaningful or data scattering. If meaningful, it implies that the previous human efforts for disaster mitigation has encountered the limitation and is now overcome by the threat of climate change.

Fig. 4.28 Number of houses damaged by rainfall disasters (data by MLIT)

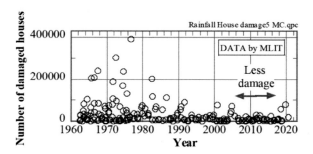

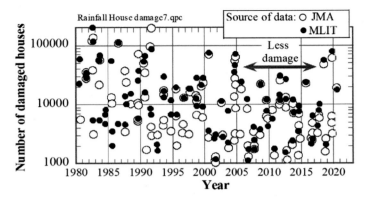

Fig. 4.29 Number of houses damaged by rainfall disasters after 1980 (data by JMA and MLIT)

4.3.3 Size of Flooded Area During Rainfall Disasters

A similar discussion as above is made of the size of flooded area during rainfall disasters. First, Fig. 4.30 compares the sizes of flooded area that have been reported by JMA and MLIT. It is noteworthy that, since 1945, JMA published the area of flooded farmland only and did not address flooded residential or urban area. Also, JMA stopped announcing the flooded area in 2007. In contrast, MLIT announces the sizes of both flooded farmland and flooded residential/urban area. However, there is no MLIT data before 1962. Figure 4.30 shows that the MLIT data is generally greater than JMA data because JMA concerns farmland only. Another reason may be, as mentioned above, different ways of data collection of the two institutions.

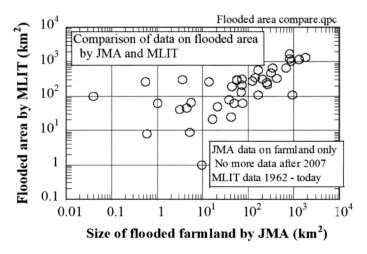

Fig. 4.30 Comparison of sizes of flooded area reported by JMA and MLIT

4 Views on Recent Rainfall-Induced Slope Disasters and Floods

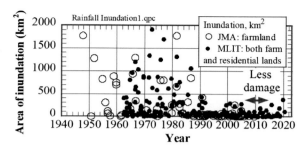

Fig. 4.31 Flooded area during rainfall disasters reported by JMA and MLIT

Figure 4.31 plots the areal size of inundation since 1940s. Both JMA and MLIT data are included. It is seen that the disasters until early 1980s were profound but the situation changed in late 1980s. This finding is consistent with that on the number of damaged houses (Figs. 4.27 and 4.28). It appears also that the inundation size was minimal in the middle of 2010s, followed by increase. This finding is similar to the discussion on house damage numbers. Figure 4.32 illustrates this trend more in detail.

As a summary of this chapter, Fig. 4.33 illustrates both number of damaged houses and size of flooded area. The greater value out of JMA and MLIT data is employed here. It is important that the size of disaster has been decreasing from 1980 to 2010s and achieved the minimal. However, this trend changed to increasing after 2016. Whether this very recent situation is significant or just data scattering, there is a fear that the long-term efforts of disaster mitigation by levees and dams are overcome by the global warming and increasing heavy rains and that the more difficult time is coming. To shed light on this issue, more discussion will be made in the next chapter.

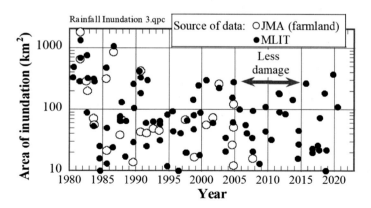

Fig. 4.32 Flooded area during rainfall disasters after 1980

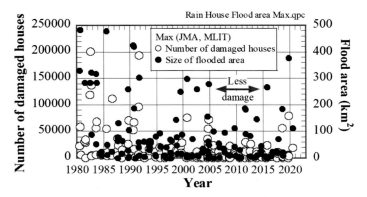

Fig. 4.33 The recent trend of house damage and flood caused by rainfall disasters

4.4 Vulnerability Indices for Rainfall-Induced Disasters

4.4.1 Significance of Vulnerability Index

The question in the previous chapter was whether or not the recent trend of damage size is the consequence of heavy rains that are becoming stronger due possibly to the global climate change. Another simple possibility is that the minimal damage size was a consequence of the lack of extremely heavy rain in mid-2010s (Fig. 4.5). To discuss this issue quantitatively, the present study attempts to use a "Vulnerability Index", VI, that is defined by

$$VI = (\text{Size of damage})/(\text{Intensity of rainfall}) \qquad (4.1)$$

where the size of damage is represented by the number of damaged houses or the size of the flooded area, while the intensity of rainfall is represented by the maximum total rainfall throughout the disaster period or daily rainfall. One-hour rainfall data may give extreme numbers despite that 1 h rainfall is not powerful enough to induce disaster. Further, JMA and MLIT give different values of damage size and the present study employs the greater number among them

$$\text{Size of damage} = \text{Max}(\text{JMAdata}, \text{MLITdata}) \qquad (4.2)$$

It is also possible that the nation's No.1 rainfall intensity is too extreme. Therefore, the other choice is the use of No.10 of the observed rainfall intensity in order to avoid the extremeness of the No. 1 record. Figure 4.34 compares the No. 1 and No. 10 of 24 h rainfall records during disasters. Approximately 30% difference is recognized between No. 1 and No. 10 rainfalls. Accordingly, there are eight types of vulnerability index as listed in Table 4.1.

4 Views on Recent Rainfall-Induced Slope Disasters and Floods

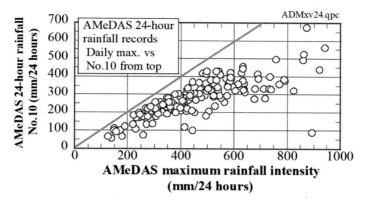

Fig. 4.34 Comparison of No.1 and No. 10 24 h rainfall records during disasters

Table 4.1 Eight types of vulnerability index, *VI*, employed in discussion

	Number of damaged houses*	Size of inundated area (km^2)*
Number 1 of total rainfall	*VI*(house #1 Total)	*VI*(area #1 Total)
No. 10 of total rainfall	*VI*(house #10 Total)	*VI*(area #10 Total)
No.1 of daily/24-h rainfall	*VI*(house #1 24 h)	*VI*(area #1 24 h)
No.10 of daily/24-h rainfall	*VI*(house #10 24 h)	*VI*(area #10 24 h)

* The number of damaged houses and the size of inundated area for *VI* calculation is the greater value out of JMA and MLIT data

4.4.2 Calculation of Vulnerability Indices Based on Number of Damaged Houses

This section studies the vulnerability indices (*VI*) on the basis of the number of damaged houses. Figure 4.35 was obtained by using the No. 1 total rainfall record during the disaster period. First, it is found that *VI* has been decreasing continuously since 1945. This is the great achievement of the efforts for disaster mitigation. The decreasing trend became obvious particularly after 1970. However, the decreasing trend is not clear after 2000 and *VI* remains around 100. In this regard, the author points out that the event in 2012 (Osaka-Kinki) might be a singular point where 23,175 houses (MLIT data) were flooded below the first floor (ground floor) (water depth being not greater than 30 or 50 cm) while only 3913 houses (MLIT data) were inundated above the floor. This big contrast of four times difference between numbers is singular as illustrated in Fig. 4.36. Furthermore, the ratio of two numbers is plotted in Fig. 4.37 where the ratio of nearly six in 2012 was out of the trend in

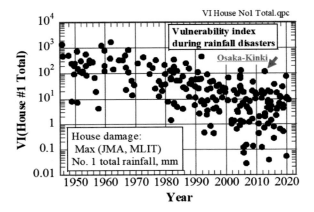

Fig. 4.35 Vulnerability index based on the number of damaged houses and No.1 total rainfall throughout the disaster period

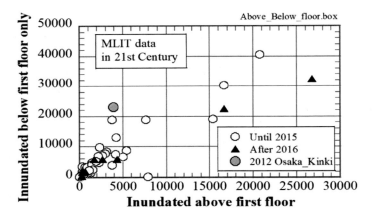

Fig. 4.36 Comparison of MLIT data on number of houses inundated above versus below the first '(ground) floor

Fig. 4.37 Temporal change of the number of houses inundated above and below the first floor (MLIT data)

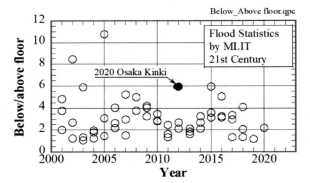

4 Views on Recent Rainfall-Induced Slope Disasters and Floods 67

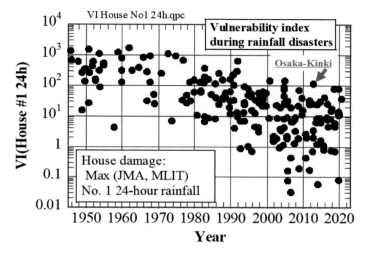

Fig. 4.38 Vulnerability index based on the number of damaged houses and No.1 24 h rainfall

mid-2010s. Because of these reasons, the author continues discussion by eliminating the 2012 Osaka-Kinki data.

Figure 4.38 shows the variation of *VI* in terms of house damage and No. 1 rainfall for the period of 24 h. This *VI* has been decreasing since 1970 but appears to be stabilized after 2000. If the data of 2012 Osaka-Kinki is eliminated, it can be said that the decreasing trend continued until mid-2010s and possibly started to increase after 2016 as suggested by two data in 2018 (typhoon No. 7 "Prapiroon" affecting Okayama, Hiroshima, and western Japan) and 2019 (East Japan during the typhoon No. 19 "Hagibis"), together with the rainfall disaster in 2020 (Kumamoto, etc. in July). Note, however, that the increasing trend in very recent years is subject to change. The recent trend of this *VI* is further examined in Fig. 4.39. By eliminating

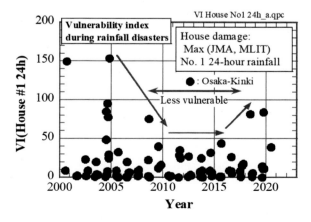

Fig. 4.39 Recent change in vulnerability index based on the number of damaged houses and No.1 24 h rainfall

the data on Osaka-Kinki, the increasing trend in very recent years is suggested. At least it is confirmed that the previous decreasing trend ceased in the first half of 2010s and the damage size is increasing under the effect of increase of rainfall intensity. Good time has gone and we should not be optimistic about safety from rainfall-induced disasters.

Figures 4.40 and 4.41 indicate *VI* calculated by No.10 s of total and 24 h rainfalls. Note that data on No. 10 rainfall is available after 1970s only. They show again the

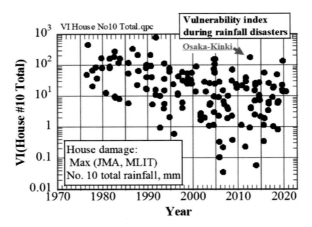

Fig. 4.40 Vulnerability index based on the number of damaged houses and No.10 total rainfall throughout the disaster period

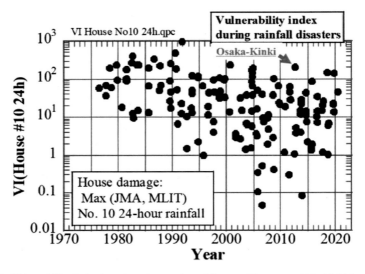

Fig. 4.41 Vulnerability index based on the number of damaged houses and No.10 24 h rainfall

4 Views on Recent Rainfall-Induced Slope Disasters and Floods

decreasing trend after 1990. By eliminating the 2012 Osaka-Kinki data, it may be said with some reserve that *VI* is increasing nowadays.

4.4.3 Calculation of Vulnerability Indices Based on Size of Inundated Area

This section addresses another type of vulnerability indices (*VI*) on the basis of the size of the flooded area. The flooded area size is the maximum between JMA and MLIT data which is denoted by Max(*JMA, MLIT*). Figure 4.42 demonstrates the change of *VI* in terms of the No.1 of the total rainfall during each disaster. After very high level until 1980, this *VI* started to decrease until 2000. After this period, *VI* is held constant, although two data in the very recent time (after 2016) may suggest increasing trend. Finding on *VI* is thus consistent with that of the number of damaged houses.

Figure 4.43 concerns the areal vulnerability in terms of No. 1 of 24 h rainfall during respective disaster event. Similar to the previous diagram, *VI* decreased since 1980 and remained constant in 2010s. Figures 4.44 and 4.45 indicate *VI* of area divided by No. 10 s of total and 24 h rainfalls. The decreasing trend until 2000 followed by stable level is seen as well.

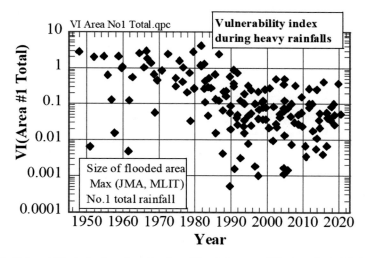

Fig. 4.42 Vulnerability index based on the size of flooded area and No.1 total rainfall throughout the disaster period

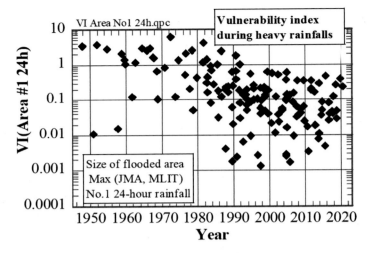

Fig. 4.43 Vulnerability index based on the flooded area and No.1 24 h rainfall

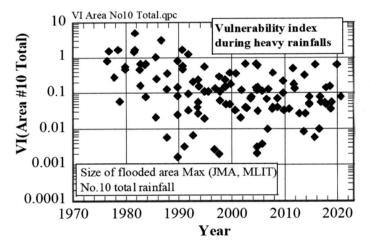

Fig. 4.44 Vulnerability index based on the size of flooded area and No.10 total rainfall throughout the disaster period

4.4.4 Trend of Vulnerability in the Twenty-First Century

One of the major concerns in this paper is the vulnerability to rainfall disaster in the recent years, because the recent trend will probably continue during the coming decades. The discussion on *VI* above showed that vulnerability decreased in the twentieth century but that the trend changed in the twenty-first century, *VI* remaining constant or increasing, depending on the choice of parameters. Because the rainfall

4 Views on Recent Rainfall-Induced Slope Disasters and Floods

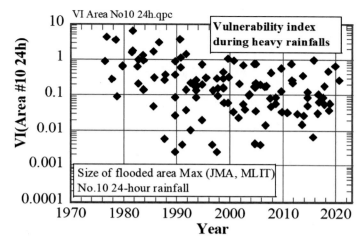

Fig. 4.45 Vulnerability index based on the flooded area and No.10 24 h rainfall

intensity is increasing in the recent years (Figs. 4.24 and 4.25), *VI* values, even if they are constant and do not increase, imply the increasing trend of the number of damaged houses and the size of the flooded area. Moreover, this situation may be aggravated if *VI* is increasing nowadays. To shed light on this issue, study is made of the data in the twenty-first century.

Figure 4.46 depicts the change of the number of damaged houses and the size of the flooded area in the twenty-first century only. The maximum of the JMA and MLIT data is used here. It is evident here that the damage upper bound achieved the minimal in the years around 2010 but started to increase thereafter. Then the vulnerability index in terms of No. 1 of 24 h rainfall is presented in Fig. 4.47. *VI* was

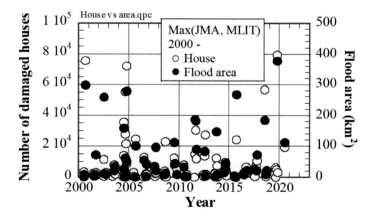

Fig. 4.46 Number of damaged houses and size of flooded area in twenty-first century

Fig. 4.47 Variation of vulnerability indices in twenty-first century

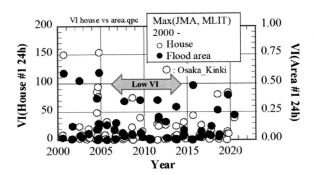

minimal in early 2010s, followed by increase. If this trend continues, it must be said that profound rainfall disasters are likely to occur from now on. The "safe" time in early 2010s is over.

4.5 Reasons for Increasing Vulnerability

As discussed above, the vulnerability index (VI) stopped decreasing at the end of the twentieth century. This is a serious problem when the chance of heavy rain increases under the possible effects of global climate change. In this regard, discussion is made on a question why VI does not decrease anymore.

The traditional approach toward disaster mitigation relied on construction of dams and retention basins as well as river levees. Obviously, a great success was attained in the twentieth century as shown by reduced VI. Practice of those measures, however, became difficult at the end of the century because, according to the author's personal impression, the community forgot the memory of big disasters and started to feel safe enough. That situation may be related with "normalcy bias" of human, meaning that people know about disasters but do not think disasters will happen to themselves. Figure 4.48 shows the site of a planned dam whose construction was called wasting of tax money and suspended. Although an alternative measure for disaster mitigation was sought for since 2008 in this river basin, no conclusion was reached for more than 10 years. Heavy rain occurred in July 2020 in this area and triggered substantial disaster in the downstream area (Fig. 4.49). This area is situated in former river channel that is now separated by levees. Accordingly, discussion is going on now what would have happened if this dam had been constructed and whether or not the dam construction project should be re-started. Thus, people's optimistic belief changes only after disaster.

Retention basin is another measure to store river water for some time and control the peak water level in the downstream area (Fig. 4.50). This was effective in July 2020 in Northeast Japan (Yamagata Prefecture). Its limitation is that implementation

4 Views on Recent Rainfall-Induced Slope Disasters and Floods 73

Fig. 4.48 Site of planned Kawabe Dam in Kumamoto Prefecture whose construction was resumed but then suspended

Fig. 4.49 Houses in Kuma Village near Hitoyoshi City damaged by the flood disaster in July 2020

needs big area and also water retention in the midstream cannot help the upstream area.

Levee is also important in flood control. Obviously, higher levees can prevent overtopping more efficiently (Fig. 4.51). The first problem is that lack of overtopping in the upstream region brings more river water to the downstream region and increases the flood risk there. The second problem is that construction of higher levee has to acquire more footprint land (Fig. 4.52a). Obviously, purchasing land over kilometers

Fig. 4.50 Okubo Retention Basin of the Mogami River that contributed to reduce the peak water flow in July 2020

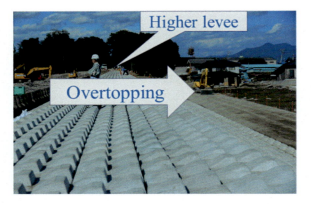

Fig. 4.51 Kinu River levee in Joso City (Fig. 4.1) that was overtopped at its lowest part in September 2015

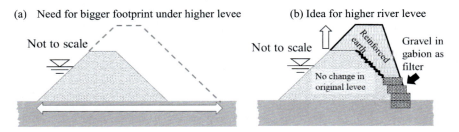

Fig. 4.52 Discussion on height of levee

(a) Condominium building in river channel   (b) Low apartment in Kawasaki, SW of Tokyo

Fig. 4.53 Residential buildings located within river channel

along river takes years or even decades. Thus, construction of river levee triggers highly political argument in the local community that may last for decades, in the extreme case, without conclusion. Dams and retention basins have the same problem. As an alternative, many people discuss about early warning and evacuation. This is a very important measure to help people's lives. However, it cannot save properties and infrastructures whose loss results in difficulties in reconstruction of damaged community.

Because traditional mitigation measures have those limitations as mentioned above, people's own efforts are extremely important. However, problems due to normalcy bias are often encountered. In the case of Fig. 4.53a, there used to be a restaurant where guests enjoyed good river view. Afterward, a condominium building was constructed and the residents refused levee in front of their place because good river view would be lost. In consequence, the basement of this building was flooded in 2019. Figure 4.53b shows another place inside levees where the basement residents do not care the risk of flood. Due to historical reasons, people are still living here. Is it reasonable to say that they should live there on their own risk?

Figure 4.54 shows a place where public safety contradicted against private land ownership. This site had a natural levee that was supposed to protect the community from flood. The landlord wished to cut the natural levee and install many solar panels there for business. Although the local community and government tried to persuade him not to do so, the natural levee was out of the legally specified river channel and the law allowed the landlord to do whatever he wished within his own property. In 2015, river water overtopped here shortly after the first overtopping at the site of Fig. 4.51. It was unfortunate that the landlord could not understand the importance of natural levee and believed that he could do on his own land whatsoever he wanted. Afterward, the national government purchased this land and constructed a new levee left half of Fig. 4.54.

People are controlled by normalcy bias and misunderstand that disaster mitigation is a business of public sectors and that they, as tax payers, do not have to worry about it for themselves. People should learn that they have to do something if they want to remain safe during natural disasters.

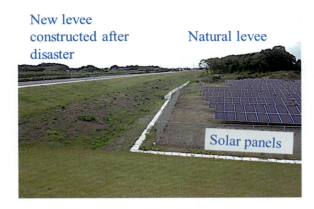

Fig. 4.54 Natural levee along the Kinu River that was removed for business of landlord

Fig. 4.55 Apartment building resting on elevated foundation

4.5.1 What to Do

The author is going to discuss his future scope in what follows. Traditional measures such as dams, retention basins, and levees should be promoted. However, they have several limitations in their present shape and, therefore, have to be improved. More details are as what follows:

- Dams are efficient but are filled with sediments with time. Removal of sediments from the reservoir is not easy but important for their longer life.

4 Views on Recent Rainfall-Induced Slope Disasters and Floods

- Retention basin is efficient. Not only the farmland but even urban area may be designated as new retention basin because small rivers in urban area are prone to flood in the recent times. If surface retention is not easy, underground retention basin is another choice, although excavation is costly.
- Levee has to be made higher without acquiring greater footprint. Figure 4.52b suggests possibility to increase the height without enlarging the horizontal size of the levee. The original part of the levee is not touched upon at all with consideration of the traditional philosophy of river engineering and the additional part is constructed by reinforced soil (e.g., fiber reinforced soil).
- The author cannot suggest any particular height of levee because, as stated before, levee construction is a highly political issue in local communities and requires years of local negotiation before engineering design.
- Slope surface is covered by concrete blocks as protection from erosion.
- If necessary, private rights of landlords have to be restricted. At least, land for disaster mitigation has to be purchased smoothly.

Self-protection of people and communities consist of the following issues:

- Those who insist on living in disaster-prone areas should do so on their own risk. The public sectors should provide them detailed information about the risk of possible flood.
- In case that relocation is not possible, residence should be uplifted (Fig. 4.55) on the cost of the owners. Construction of private levee is another choice.
- Warning and evacuation are essential in case of emergency. Note that it is difficult for warning and evacuation to save properties in case of emergency.

4.6 Conclusion

The present study attempts to quantitively foresee the risk of rainfall-induced disasters by examining the relationship between the extent of rainfall-induced disasters and the intensity of rainfall. Data from 1945 to 2020 was collected from literatures as well as publications of Japan Meteorological Agency (JMA) and the Ministry of Land, Infrastructure, Transport and Tourism (MLIT). Therein, care was taken of the current situation of global warming and climate change that may affect the rainfall intensity and disaster extent. The major points and conclusions in this study are as what follows:

(1) The extent of damage is quantitatively expressed either by the number of damaged houses or the size of flooded area. These two types of data are easily available in publications throughout many decades.
(2) Rainfall intensity is continuously increasing in the past decades, while the total annual precipitation remains unchanged.
(3) On a long-term basis, temperature has been rising as well. However, the data for this comes from lake freezing, melting of ice that do not account for summer weather.

(4) Sea level rise does not yet affect our community.
(5) The damage indices, which are the number of damaged houses and the size of flooded area, were very high from 1945 to 1970s, decreased from 1980s to 2010, and remained minimal afterward. There is a possibility that the damage indices appear to be increasing since 2016, but it is too early to confirm this issue.
(6) In the worst scenario, the increasing trend suggests that the climate change increases the intensity of heavy rain, overcomes the human efforts of disaster mitigation, and thereby aggravates the disaster size. It is, however, too early to draw a final conclusion on this point.
(7) A set of vulnerability indices (*VI*) were defined as the ratio of damage size/rainfall intensity.
(8) *VI* decreased from very high values before 1970s and became minimal in 2010s. Some data suggests its increase in very recent years but it is too early to draw a final conclusion on this issue as stated above.
(9) Whether *VI* is constant or increasing nowadays, the increase in rainfall intensity will worsen the induced damage. Thus, good time has gone.
(10) Disaster mitigation is not a full responsibility of public sectors. People's own efforts for safety will be essential from now on because conventional mitigation measure has reached limitations and damage size will be aggravated.

References

AMeDAS by Japan Meteorological Agency. https://www.data.jma.go.jp/obd/stats/etrn/
Arakawa H (1954) Fujiwara on five centuries of freezing dates of Lake Suwa in the Central Japan, Archiv für Meteorologie, Geophysik und Bioklimatologie. Serie B (Theor Appl Climatol) 6(1):152–166
Banglapedia. http://en.banglapedia.org/index.php?title=Cyclone
Carbognin L, Teatini P, Tosi L (2009) The impact of relative Sea Level Rise on The Northern Adriatic Sea Coast, Italy, Management of Natural Resources, sustainable Development and Ecological Hazards II. WIT Trans Ecol Environ 127(12):137–148
Cittá di Venezia (2019) Le acqua alte eccezionali. https://www.comune.venezia.it/it/content/le-acque-alte-eccezionali
Fujiwara S, Arakawa H (1954) Five centuries of freezing dates of Lake SUWA (36°N, 138°E) in the Central Japan. J Meteorol Res 6(5):127–137 (In Japanese) (1954)
Grinsted A, Ditlevsen P, Christensen JH (2019) Normalized US hurricane damage estimates using area of total destruction, 1900–2018. Proc Natl Acad Sci 116(48):23942–23946
Hobbs WH (1911) Requisite conditions for the formation of ice ramparts. J Geol 19(2):157–160
Ishiguro N, Kajiwara M, Fujita T, Akiba Y, Touchart L (2002) Heterogeneity of the Omiwatari records of Lake Suwa as the database for winter temperature estimation. Internationale Vereinigung Für Theoretische Und Angewandte Limnologie: Verhandlungen 28(2):1107–1110
Kajander J (1993) Methodological aspects on river cryophenology exemplified by a tricentennial break-up time series from Tornio. Geophysica 29(1–2):73–95
Lee HS (2013) Estimation of extreme sea levels along the Bangladesh coast due to storm surge and sea level rise using EEMD and EVA. J Geophys Res: Oceans 118(9):4273–4285

MLIT (2008) Ministry of Land, Infrastructure, Transport and Tourism: climate change induced by global warming. In: Presented at 7th meeting of Special Committee on mitigation of extreme flood disaster (in Japanese)

NASA Global temperature (2020). https://climate.nasa.gov/vital-signs/global-temperature/

Sharma S, Magnuson JJ, Batt RD, Korhonen J, Aono Y (2016) Direct observations of ice seasonality reveal changes in climate over the past 320–570 years. Nat Sci Rep 6:1–11

Suenaga M, Matsumoto H, Itabashi N, Mihara M, Umeki Y, Isobe M (2003) Anomaly in tidal level in Hiroshima Bay. In: Proceedings of the coastal engineering, JSCE, vol 50, pp 1316–1320 (in Japanese)

Sugimoto Y, Fujii S, Moriya T, Sasatani T (1981) Observations of ice-rampart and icequake activity in Lake Kussharo. Geophys Bull Hokkaido Univ 40:79–91 (in Japanese)

Tanaka A (1918) Limnological study on Lake Suwa, publ. Iwanami, p 681 (in Japanese)

Tanaya T, Tokoro T, Watabe Y, Kuwae T (2018) Field measurements and analyses of carbonate production by a coral reef ecosystem : towards the low-water line protection of remote islands. Rep Port Airport Res Inst 57(2): 3–33 (in Japanese)

Towhata I, Goto S, Goto S, Akima T, Tanaka J, Uchimura T, Wang G, Yamaguchi H, Aoyama S (2021) Mechanism and future risk of slope instability induced by extreme rain-fall event in Izu Oshima Island, Japan. Nat Disasters 105(1):501–530

Toyota T (2011) Natural disaster in tourist spot of Miyajima with cultural heritage. Historical disaster studies in Kyoto, 12:9–21 (in Japanese)

Umeki Y (2003) On the state of occurrence and estimation for factors of occurrence of extra-high tide events recently observed in western Japan. In: Proceedings of the Coastal Development Institute of Technology 3:5–8 (in Japanese)

Van den Dool HM, Krijnen HJ, Schuurmans CJE (1978) Average winter temperatures at De Bilt (the Netherlands): 1634–1977. Clim Change 1(4):319–330

Yoneyama K (1988) Ice rampart in Lake Suwa. Series "Dialogues on Tenryu River", publ. MLIT Tenryu River Upstream Office, vol 10 (in Japanese)

Chapter 5
Appropriate Technology for Landslide and Debris Flow Mitigation in Thailand

Suttisak Soralump and Shraddha Dhungana

5.1 Introduction

Extreme weather and climate events have increased in frequency and are projected to continue increasing in this century (Seneviratne et al. 2012). These events can impact humans and ecosystem extremely and include major destruction of assets, loss of human lives, and loss of and impacts on plants, animals, and ecosystem services (Handmer et al. 2012; Miura and Nagai 2020). Landslides are common geomorphic events on fragile, steep slopes of the mountains in Thailand. Recently, the frequency of rain-triggered landslides in Thailand has increased and has gained momentum, coincident with the effects of climate change (Fig. 5.1). The northern and southern part of Thailand is the most vulnerable part of the country subjected to landslide hazard (Fowze et al. 2012; Soralump 2010).

Compared to other natural hazards (e.g., floods, earthquakes, and storms), the impact of landslides is often underestimated because the affected areas are mostly on a local scale (Kalia 2018). Landslide hazard are expected to increase in the future through population growth, new settlements in landslide-prone areas, and climate change (Gariano and Guzzetti 2016). Although the occurrence of future landslides cannot be prevented, the magnitude of impact in terms of loss of life and destruction of property can be kept within reasonable bounds through preventive and mitigation measures (Tanavud et al. 2000).

S. Soralump (✉) · S. Dhungana
Geotechnical Engineering Research and Development Center, Kasetsart University, Bangkok, Thailand
e-mail: soralump_s@yahoo.com; fengsus@ku.ac.th

© The Author(s), under exclusive license to Springer Nature Singapore Pte Ltd. 2023
H. Hazarika et al. (eds.), *Sustainable Geo-Technologies for Climate Change Adaptation*, Springer Transactions in Civil and Environmental Engineering,
https://doi.org/10.1007/978-981-19-4074-3_5

Fig. 5.1 Landslide at Nan Province, Thailand

5.2 Types of Landslides

A landslide is the movement of a mass of rock, earth, or debris down a slope (Cruden 1991). In landslide classification, there are great difficulties due to the fact that phenomena are not perfectly repeatable; usually being characterized by different causes, movements, and morphology, and involving genetically different materials. For this reason, landslide classifications are based on different discriminating factors, sometimes very subjective (Souza et al. 2016). However, depending on the size of slide mass and potential for loss of life, landslide has been classified into three categories.

5.2.1 Slope Failure

Slope failure is a phenomenon that occurs due to the weakened self-stability of the earth under the influence of various factors like rainfall or an earthquake. The sudden collapse of slope near the residential area results in the loss of life and property. Mostly slope failure occurs from cut-slopes without an adequate protection (Fig. 5.2). The failure mass is limited and mostly slides down to replace the soil mass that has been cut (Fig. 5.3).

Fig. 5.2 Slope failure in cut slope

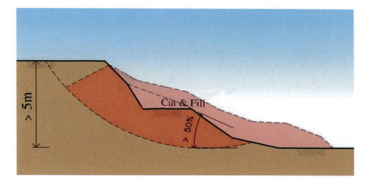

Fig. 5.3 Graphical representation of slope failure in cut slope

5.2.2 Landslide

Landslide involves a large area composed of several slope failure and movement. It affects the part or whole of the hill slope area. Most of the cases of landslide are related with the increase in groundwater table from rainfall infiltration. Doi Chang village in the northern region of Thailand is a good example of landslide (Soralump 2017) where the whole village has been moving with slow rate resulting in the destruction of house and infrastructure (Fig. 5.4).

Fig. 5.4 Observed movement of house and infrastructure in Doi Chang village, Chaing Rai, Thailand

5.2.3 Debris Flow

Lastly, the most fatal type is debris flow. They develop when hill slope is subjected to heavy rainfall and mostly in case of extreme precipitation events. Debris flow occurs when the flood debris flow down the flow channel and inundate the downhill area. In 2011, numerous villages were swept away by landslide and debris flow at Khao Phanom Benja National Park of Krabi Province in the southern part of Thailand (Fig. 5.5) (Soralump 2010; Iyaruk et al. 2019).

Fig. 5.5 Landslide and Debris Flow at Krabi Province, Thailand

5.3 General Landslide Mitigation Scheme

A hazard doesn't necessarily have to become a disaster, or can be minimized, if we are well prepared. Several mitigation measures have been proposed by various authors for landslide and debris flow (Popescu and Sasahara 2009). This mitigation requires good information and tools such as hazard and risk map, rainfall prediction system, sensors, and warning tools. A technology is deemed to be appropriate when it is consistent with the cultural, social, economic, and political context of each society and country. Some of the effective landslide mitigation measures are listed below.

Relocate the population living in the risk areas: This is the most effective and difficult mitigation scheme especially in the mountainous area where suitable location for settlement is hard to find. New legal regulations cannot be applied to the people who have been residing at those areas before the law was issued. Moreover, this mitigation is suitable for the area where land movement is active or where landslide and debris flow had happened in the past even in some geological period.

Warning and evacuation: The sequential steps of the evacuation process would be detection, evaluation and prediction, decision and warning to bring population at risk out of the potential hazard area. However, evacuation drill needs to be carried out periodically for key success.

Structural measures: These measures are mostly used for preventing slope failure by means of slope reinforcement or slope protection to reduce the speed of debris flow by using check dam, debris flow net, etc. This method, however, is not economically feasible for a large area.

Law, code and standard: Effective law, code, and standards are essential part of a long-term strategy and sustainability but are normally time consuming. Law can enforce all three mitigation schemes mentioned above but its effectiveness depends on the authorized representative.

5.4 Country Context that Affects Mitigation

5.4.1 Right of Landowner

In Thailand and many southeast ASEAN countries, people living in landslide risk area cannot fully clarify the ownership of the land. This is because human settlement is mostly located near the hill slope or connecting area between mountain and flat area, which has easier access to woods for hunting and collecting timber products with some flat area for housing, agriculture, or farmland. People living there for centuries get affected by the frequently occurring landslides and have adjusted their housing location accordingly. In later years, government has issued forestry conservation law to conserve the forest in the mountainous area. Village located inside the conservation area cannot verify if their village or houses have been there before or after the issuance of the law. This has created a conflict and dispute between the people and the government which has been trying to sort out this issue for years. The conflict worsens in case of landslide-prone areas. People are expected to move out to the new safe area but as their land ownership is not yet clear, government cannot issue a new land for them.

5.4.2 Law Structure and Governance

The law in Thailand has different levels starting from the constitution, act, ministerial regulations, ministry announcement, code, and standard. Laws relating with the landslide are building code act, cut and fill of earth act, land development act, disaster reduction and prevention act, etc.

Department of Disaster Prevention and Mitigation (DDPM), established in 2002, is a central state agency created under Ministry of Interior (MOI) with the responsibility to oversee the administration of disaster management responsibilities in Thailand. The national disaster management system is made up of multiple agencies and committees to carry out disaster preparedness and response activities. The Disaster Management System based on the Disaster Prevention and Mitigation Act 2007 (DPM Act 2007), came into force on 6 Nov 2007 and implements Thailand's national Disaster Management (DM) Institutional arrangement. All disaster management activities are directed and controlled by the Commander/Director at different levels; National, Provincial, District, and Local (THAILAND 2018).

In case of Thailand, flood is the most catastrophic disaster; therefore, most of the existing laws are developed to address this issue as compared to landslide. Since, severe landslides were observed in Thailand after the onset of development in late 1980s, proper guidelines for landslide mitigation have not been developed. Additionally, lack of guidance on how governmental agencies and other stakeholders should coordinate their work has been a major challenge to Thailand's disaster management system.

5.4.3 Economics

Economics has always been a major concern in mitigation itself because people from landslide-prone areas always express their uncertainty about the cost and finances after the relocation. One of the best examples would be Doi Chang village, Mae Suai District, Chiang Rai Province of Thailand. This village has been continuously moving for years and is constantly monitored by Geotechnical Engineering Research and Development Center (GERD), Kasetsart University. The area is a national reserved forest and residents there has no documentation to live or farm in the area. Because of the cool temperature and the area being 1,000 m above sea level, the conditions are ideal for Arabica coffee plantation. This yields high quality aromatic coffee making it the number one choice for coffee drinkers around the world and the attraction place for tourist. People living there rely on coffee farming and therefore are not willing to relocate as economics and other insecurities after the relocation influences people's decision of moving to a new safe place.

5.5 Landslide and Debris Flow Mitigation in Thailand

5.5.1 Multiway Warning System

Establishing an effective early warning system is a vital tool for mitigating the impacts of disaster. The more time we have, the safer people are. An effective warning system is based on high quality, real-time data with parameters having higher accuracy. In case of landslide and debris flow, precipitation is the most important parameter that needs to be considered. Various landslide prediction models (de Meij et al. 2009) can predict the amount of precipitation in advance for 3–4 days with some certain accuracy. This information can be used for advance warning. Multiway warning system, which has been developed in Thailand, uses rainfall precipitation prediction to warn the village in case of heavy rainfall (Fig. 5.6). There are 54 provinces and more than 5000 villages prone to landslide and debris flow in Thailand. It is not possible to monitor all the villages by central government and there remains a requirement for early warning system. Multiway warning system can analyze 3–4 days landslide potential area in advance using Antecedent Precipitation Model (AP model) (Thowiwat and Soralump 2010; Soralump 2009; Setpeng et al. 2020). This model analyzes rainfall threshold by using information of the precipitation that caused landslide in the past. Warning message will be given to the villages if there is probability of predicted rainfall exceeding the rainfall threshold. The accuracy of the warning from AP model, even though not high but is good enough to alert people in the prone area.

It is necessary to establish community-based warning together with the early warning using AP model. Necessary tools, knowledge, and evacuation plans are required for letting the people in the risk area respond accordingly. Once the warning

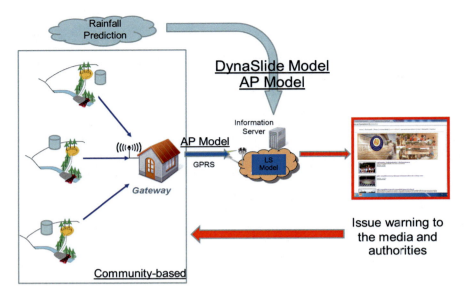

Fig. 5.6 Multiway warning system using AP model (Fowze et al. 2012)

is issued, people in the risk zone should record the accumulated rainfall precipitation from rain gauge and plan for the evacuation if rainfall exceeds the threshold. Figure 5.7 shows the timeline before and after the occurrence of landslide. Different types of detection and warning systems are placed based on this timeline. The accuracy of the warning message from prediction might be less but is enough for the preparation and evacuation while the accuracy of warning from direct measurement of land movement is high but the warning period is generally less. This concept has been used in Thailand for more than 10 years and has successfully saved many lives.

There are several ways of landslide warning and the timeline of warning being used as shown in Fig. 5.5. People of local agencies want an accurate warning system so that the false alarm or false warning would be minimized. It is better if we can get the warning faster but the accuracy in these cases will be lower but this has to be accepted for the safety of the people. Even though the accuracy of warning message is poor, if we train the local people to understand and get used to the warning and appropriate response, it will be safer. On the other hand, when we have more data and hazard information, the warning time can be reduced. Likewise, when we use instrument to detect the slide, the accuracy of warning will be high and warning time will be very low. Figure 5.7 shows the accuracy of landslide prediction and warning time based on storm, rainfall, and instruments installed. The blocks presented signifies the mode of accuracy and type of message.

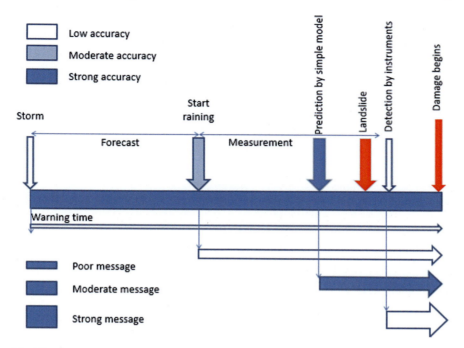

Fig. 5.7 Timeline of landslide and warning

5.5.2 Rainfall Threshold

AP model is used for landslide forecasting and is based on statistical data of rainfall precipitation that had caused landslide in the past but rainfall threshold is based on different zones of Thailand as shown in Fig. 5.8 (Kanjanakul et al. 2016). Landslide threshold can be estimated based on the plot between rainfall intensity (mm/day) and accumulated precipitation as shown in Fig. 5.9. The plot shows the cumulative rainfall for 3 days obtained from various stations of Thailand and rainfall intensity. It has been found that 3 days accumulation period of precipitation is appropriate for making landslide threshold. Initially, this model was used for local warning based on data from local rain gauge and was effective for warning 24 h prior to the landslide. Later, 3–4 days precipitation forecasting data was used for early warning of the landslide using AP model (Fig. 5.10). The accuracy of this model is further being improved using Receiver Operating Characteristic (ROC) method (Cantarino et al. 2018; Vakhshoori and Zare 2018) and comparing statistical landslide data in landslide susceptible areas. However, besides the limitation of the accuracy in forecasting precipitation, the grid size of rainfall data is 4×4 km, which may post some limited accuracy in terms of warning area.

Fig. 5.8 Rainfall threshold for different zones in Thailand

5.5.3 Dynamic Landslide Hazard Mapping

In order to warn people accurately in time with the appropriate warning area, a geotechnical model, DynaSlide, was established based on infinite slope analysis. The model can produce landslide hazard map that can be changed according to the spatial rainfall data. The analysis is coupled between infiltration analysis through unsaturated soil and infinite slope stability analysis (Fig. 5.11). Infinite slope stability analysis is performed based on effective stress analysis. The analysis is based on more than 500 undisturbed soil samples, which were collected in the period of 15 years throughout Thailand. Analysis is performed through GIS with data grid cell of 30 × 30 m. Rainfall forecasting is used as a input data and the factor of safety of slope of each grid cell (30 × 30 m) will be calculated based on this parameter. Even though, this model is more accurate, it takes much of a computer time while processing the larger area at one time. Therefore, this model is used together with AP model. AP model gives a potential area of landslide with limited accuracy and DynaSlide will analyze the particular area with finer grid for more accuracy.

5 Appropriate Technology for Landslide …

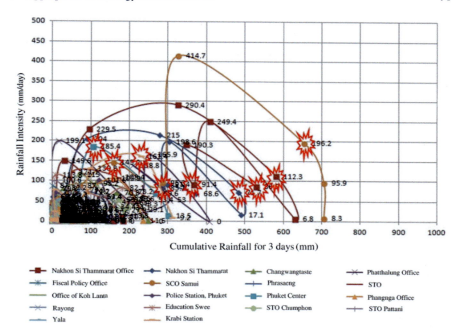

Fig. 5.9 Verification of critical API value (Soralump 2010)

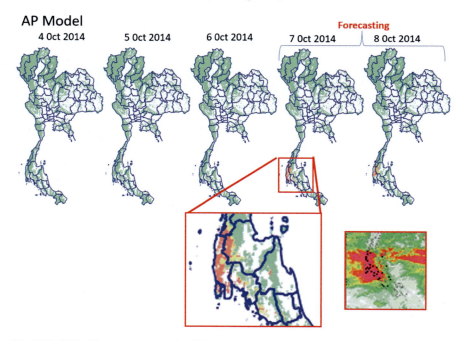

Fig. 5.10 Early Warning system using AP model

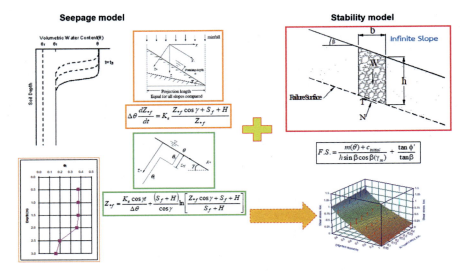

Fig. 5.11 Infiltration and slope stability analysis (Kanjanakul et al. 2016)

5.5.4 Community-Based Landslide Warning

Important key of landslide and debris flow warning in Thailand is the potential victim themselves. As mentioned earlier, there are more than 5000 villages in Thailand located within landslide-prone areas and single command operation alone cannot keep eyes on every village. Instead, people living in the prone areas should be trained for early warning. Three key components for community-based landslide warning are.

Knowledge

Set of knowledge about landslide management should be transferred to the local people. On the other hand, local information, knowledge, wisdom, and experience needs to be combined to come up with the specific landslide management of that village. Regular training is necessary to keep up through all generations in the village. Figure 5.12 shows the training conducted in Khao Phanom District, Krabi Province to transfer the knowledge regarding the early warning system installed in the village. The system has now been handed over to the local community and is being operated by local government with constant supervision from Geotechnical Engineering Research and Development Center.

Tools

Appropriate tools that are easy to operate and maintain but relevant for warning is needed to be provided to local people. At least two types of sensors, i.e., rain gauge and debris flow detection system should be installed on the mountain where landslide and debris flow tend to occur (Fig. 5.13). The signal from sensor will be sent through

Fig. 5.12 Training to local community regarding landslide warning system in Krabi, Thailand

radio system because mobile phone system cannot be relied on during heavy storm. The receiver or master station will receive information from sensors and analyze if the warning needs to be issued or not. Precipitation data from rain gauge will be analyzed according to AP model and indicate the possibility of landslide using various shades of green, yellow, and red. The shade color indicates that people can make their own decision while solid light indicates that people should consider the signal seriously and evacuate the area (Fig. 5.14). It is also important for people in the risk area to consider the surrounding information such as color of water in stream, etc. to make their own decision.

Corporation and communication

Corporation and communication among local people, local officers, and central command unit is essential for the proper implementation of community-based landslide warning system. The best way is to organize evacuation drill to review emergency preparedness and action plan (EAP and EPP) which also initiates the communication inside the village for better preparation. EAP and EPP shall consist of a clear definition of the overall structure with specific responsibilities of all the key personnel and all the units should continuously coordinate with each other.

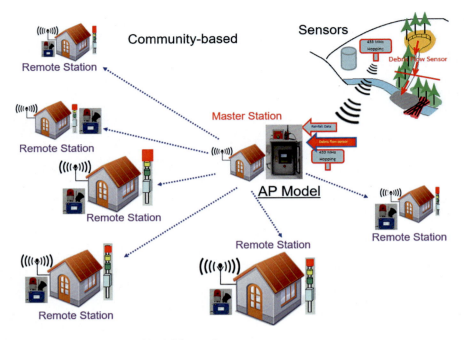

Fig. 5.13 Community-based landslide warning system

Fig. 5.14 Siren for landslide warning

Fig. 5.15 Triangle of success for landslide management system

5.6 Triangle of Success

Knowledge, Society acceptance, and Policymaker represent the three most important ingredients to ensure success. This is a framework proposed for conceptualizing the relationship between the factors that influence the efficiency of disaster management system. Triangle of success represents the policy for proper installation, maintenance, and effectiveness of landslide management system as shown in Fig. 5.15.

Knowledge: Disaster management system requires knowledge for decision-making and coordinated action. Knowledge of people in risk area and government agencies along with other stakeholders must be shared for the successful planning.

Society acceptance: People in risk area and government agencies along with other stakeholders need to have discussion before making any decisions such as issuing a regulation or standard of practice.

Policymaker: The formulation and implementation of a national policy involves information from different fields at different levels, with the active participation of each and every stakeholder. If it is to be performed effectively, the efforts from all concerned groups must be coordinated.

5.7 Conclusions

Landslide mitigation is an arduous task but various mitigation measures have been developed in Thailand to minimize the loss of life and property from landslide. It is one of the most difficult tasks to relocate the people from risk zone due to legal hindrances in Thailand but new laws are now being formulated to overcome these issues in future by coordinating with various government and non-governmental agencies for disaster mitigation. Likewise, the accuracy of Antecedent Precipitation Model (AP model) is also being improved for effective multiway warning system. This will help in improving the accuracy of landslide early warning system of Thailand. The new landslide and precipitation data are also constantly being monitored and recorded to further improve its accuracy in the future. Community-based landslide warning system has been installed in various parts of Thailand and has been handed over to the local community. Various training programs are being conducted for transferring the knowledge to the people and community living in landslide-prone areas. Emergency Action Plan (EAP) and Emergency Preparedness Plan (EPP) has been conducted in local community for preparation in case of emergency. The installation and monitoring of community-based landslide warning system will be done in various parts of Thailand in the future as well.

References

Cantarino I et al (2018) A ROC analysis-based classification method for landslide susceptibility maps. Landslides 16(2):265–282

Cruden DM (1991) A simple definition of landslide. Int Assoc Eng Geol

de Meij A et al (2009) The impact of MM5 and WRF meteorology over complex terrain on CHIMERE model calculations. Atmosp Chem Phys Eur Geosci Union (17):6611–6632. https://doi.org/10.5194/acp-9-6611-2009.ineris-00961939

Fowze JSM et al (2012) Rain-triggered landslide hazards and mitigation measures in Thailand: from research to practice. Geotext Geomembr 30:50–64

Gariano SL, Guzzetti F (2016) Landslides in a changing climate. Earth Sci Rev 162:227–252

Handmer J et al (2012) Changes in impacts of climate extremes: human systems and ecosystems. In: Managing the risks of extreme events and disasters to advance climate change adaptation

Iyaruk A, Phien-wej N, Giao PH (2019) Landslides and debris flows at Khao Phanom Benja, Krabi, Southern Thailand. Int J Geomate 16(53)

Kalia AC (2018) Classification of landslide activity on a regional scale using persistent scatterer interferometry at the Moselle Valley (Germany). Remote Sens 10(12)

Kanjanakul C, Chub-uppakarn T, Chalermyanont T (2016) Rainfall thresholds for landslide early warning system in Nakhon Si Thammarat. Arab J Geosci 9(11)

Miura T, Nagai S (2020) Landslide detection with Himawari-8 geostationary satellite data: a case study of a torrential rain event in Kyushu, Japan. Remote Sens 12(11)

Popescu ME, Sasahara K (2009) Engineering measures for landslidedisaster mitigation. In: Sassa K, Canuti P (eds) Landslides—Disaster risk reduction. Springer, Berlin, Heidelberg. https://doi.org/10.1007/978-3-540-69970-5_32

Seneviratne SI et al (2012) Changes in climate extremes and their impacts on the natural physical environment. In: Managing the risks of extreme events and disasters to advance climate change adaptation

Setpeng S, Chaithong T, Soralump S (2020) Accuracy assessment of antecedent precipitation model (AP-model) for landslide early warning system. In: The 25th National convention on Civil Engineering, Chonburi, Thailand

Soralump S (2009) Landslide risk management in Thailand using API model. In: Geotechnical infrastructure asset management, 2009. EIT-JSCE international symposium 2009. Imperial Queen's Park Hotel, Bangkok, Thailand, Organized by EIT, JSCE, Kyoto University, AIT

Soralump S (2010) Rainfall-triggered landslide: from research to mitigation practice in Thailand. Geotech Eng J SEAGS & AGSSEA 41(1). ISSN 0046-5828

Soralump S (2017) A Study on the failure behavior of colluvium soil slope: a case of Doi Chang village. In: The 22nd National Convention on Civil Engineering. 2017: Nakhon Ratchasima, Thailand

Souza et al (2016) Case study and forensic investigation of landslide at Mardol in Goa. In: 5th International conference on forensic geotechnical engineering 2016, Bengaluru

Tanavud C et al (2000) Application of GIS and remote sensing for landslide disaster management in Southern Thailand. J Nat Disaster Sci 22:67–74

THAILAND (2018) Disaster management reference handbook: center for excellence in disaster management and Humanitarian Assistance (CFE-DM)

Thowiwat W, Soralump S (2010) Critical API model for landslide warning. In: 15th National Civil Engineering conference, Bangkok, Thailand

Vakhshoori V, Zare M (2018) Is the ROC curve a reliable tool to compare the validity of landslide susceptibility maps? Geomat Nat Haz Risk 9(1):249–266

Chapter 6
Slope Creep Instability in Krajang Lor Village, Magelang Regency, Central Java, Indonesia: Inducement and Developmental Prediction

Tran Thi Thanh Thuy

6.1 Introduction

Soil creep, the imperceptibly slow, steady downward movement of soil mass resulting from viscous shear stresses or gravitational stresses that produce permanent deformation but are insufficient to cause shear failure (Ziemer 1977) is a common process on expansive soil slopes (Lytton et al. 1980) with creep rate less than a few mm per year (Fleming and Johnson 1975; Owens 1967; Soma et al. 1992). Although few publications regarding to creep in volcanic residual soil slope, a creep body had been observed on such that soil slope at the Krajan Lor Village Salaman District, Magelang Regency, Central Java Province, Indonesia were underlain by volcanic breccia and lava (Rahardjo et al. 1995). A number of wide tension cracks were observed on the slope surface, inducing crack on the walls of local houses. The soil cracks have been threatening approximately 200 houses in the Village and at least 43 houses have been damaged due to cracks in the walls.

According to (Lytton et al. 1980), the different causes may lead to the different mechanisms of soil creep, but in general, it is governed by controlling factors such as mineralogical composition, change in water content, driving stress, shear strength parameters (i.e., the friction angle and the cohesion), slope inclination, and thickness of the active creep zone. In this paper, mechanism of volcanic soil creep paper research were: (1) to investigate the mains factor inducing soil creep; (2) to develop slope stability model based on Mohr-Coulomb failure criterion that allows evaluate the thickness of creep zone and critical slope angles; (3) to develop logarithmic creep modeling to predict developmental creep displacement. The outcome of this study,

T. T. T. Thuy (✉)
Project Management Unit of Construction Investment, People's Committee of District 12, Ho Chi Minh City, Viet Nam
e-mail: t4.thanhthuy@gmail.com

is, then, utilized as one of references for local planning of mitigation works and land uses.

6.2 Soil Creep Intensity

Krajan Lor Village is situated on the plateau of Magelang Regency, Central Java, Indonesia (Fig. 6.1) covering about 9.08 hectares of housing area and 24.7 ha of agriculture area. A slowly creeping body was identified in the 9.08 ha-housing area where concentrates dense houses, high concentrated population, and livestock (Fig. 6.2). The creep slope has a high inclination at the upper part (i.e., mean of 30°) and a gentle inclination at the lower part (i.e., means of 13°).

Initiation of soil creep had been observed firstly by local people since 2006. After a decade, accumulated creep results in benched trees and tension cracks (Fig. 6.3). Based on land uses map (Fig. 6.2), those cracks distribute in two major areas: (1) The top of the slope was characterized by a local road crossing along. A main crack with approximately 134 m in length and approximately 21 mm in width locates on the pavement right at the middle of the road. Additional, there were a number of minor cracks or ripples that have the length reaching to approximately 50 m and the width reaching to 2 mm. (2) The middle of the slope was characterized by high concentration of housing. Cracks occurred on the walls of 44 houses some of those have the cracks width reaching to 40 mm. Most of the damaged houses were in parallel lines. The cross section of the slope is illustrated in Fig. 6.4. Based on the results, there are three recognizable soil units:

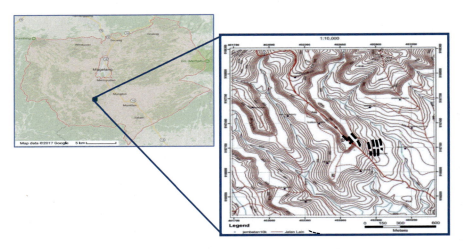

Fig. 6.1 Location of Krajang Lor Village, Salaman, Magelang, Indonesia

6 Slope Creep Instability in Krajang Lor Village …

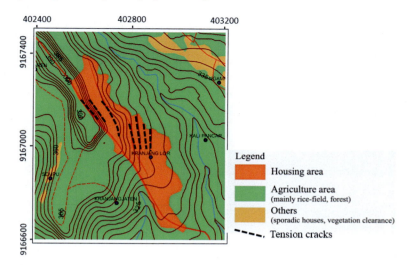

Fig. 6.2 Land uses map of Krajang Lor Village and surroundings area

Fig. 6.3 Evidences of soil creep on hill slope in Krajang Lor Village. **a** Benched trees. **b** Crack on the housing wall. **c** Crack on the ground surface. **d** Crack con the pavement of main road

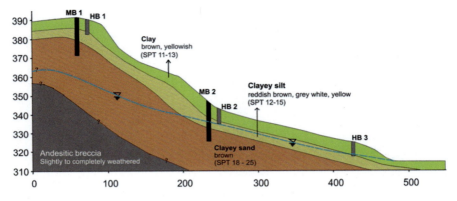

Fig. 6.4 Illustration of cross section between MB1 and MB2

(1) *Clay*: brown, yellowish, and very plastic. This is the top layer distributing from the soil surface to a depth of approximately 4.5 m. The SPT value is from 11 to 13.

(2) *Clayey silt*: reddish brown, gray white, yellow. A thin clayey silt layer is overlain by the top clay layer and distributes from the depth of approximately 4.5–10.5 m. The SPT value is from 12 to 15.

(3) *Clayey sand*: brown, fine to medium. The clayey sand layer locates beneath the clay layer with unknown thickness because the bottom of bore holes is in this layer. The SPT value is from 18 to 25. Some sandstone fragments which are moderate compact and moderate hard have been found indicating that this sand layer is weathering result of sandstone.

Generally, the Krajang Lor Village and surrounding areas were composed of mainly volcanic residual soil that showed a homogeneous spatial distribution of lithology. On the other hand, soil creep occurred mostly on the volcanic residual soil. Hence, the Krajang Lor Village and surrounding areas had almost the same lithological condition. But soil creep took place only on the housing area of Krajang Lor Village while absence of creep in the rest areas, Therefore, there must be important factors that might exist only on the housing area and made the housing area differ from other areas.

6.3 Methodology

6.3.1 Field Investigation and Laboratory Testing

Field investigation: Drilling works were conducted including 3 hand-drilling boreholes up to 6 m depth and 2 mechanical-drilling boreholes up to 20 m depth. undisturbed soil samples, which were 90 mm in diameter and 600 mm in height, was taken with depth interval of 1.5 m and 4.5 m depth for hand-drilling and mechanical-drilling bore holes, respectively. During mechanical drilling, SPT were carried out in every borehole at intervals of 2 m.

Laboratory testing: All the samples were brought to engineering properties testing while 3 representative soil samples collected at depth of 1.5–2.0 m of 3 hand-drilling bore holes were brought to XRD and SEM analyses. Moreover, series of consolidated drained (CD) triaxial tests and drained deviatoric triaxial creep test were conducted to identify stress-strain relation and deviatoric creep behavior of soil sample, respectively.

Consolidated drained (CD) triaxial test: The soil specimens were cylindrical in shape with approximately 50 mm in diameter and 100 mm in height. During testing, the confining pressures of 40 kN/m^2 were used in order to illustrate the in-situ condition as soil profile at the depth of 2.5–30 m. Strain rate performed in the triaxial tests was 1%/min. Pure water was used as cell fluid at the laboratory temperature.

Drained deviatoric triaxial creep test: A series of triaxial creep tests was conducted by applying constant deviatoric stresses ($\sigma_1-\sigma_3$) ranging from 20 to 90% of the soil shear strength) to natural soil specimens. Drained triaxial creep test with confining pressure of 40 kPa was conducted for a duration maximum of 6.10^5 s, but it could be finished earlier in case the sample was a rupture.

6.3.2 Slope Stability Modeling Based on Mohr-Coulomb Failure Criterion

The soil slope medium was considered likely the slope illustrated model as in Fig. 6.5 with slope angle β. Select randomly a location named A on the soil surface, there would be parameters a (i.e., the length from A to the toe of the soil slope) and z

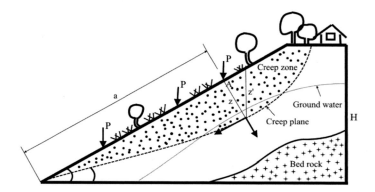

Fig. 6.5 Slope illustrated model

(i.e., the thickness of the object considered as the soil creep zone). Hence, the object rested on a creep plane inclined with angle α.

A slope stability model based on Mohr-Coulomb failure criterion is proposed by combining slope geometry parameters with the Coulomb yield strength function. By substituting normal stress, σ, and shear stress, τ, two components derived from gravitational stress, into the Coulomb straight-line equation, $\tau_r = c_r + \sigma \tan\phi_r$ (Robert 2006), we have

$$\left(\gamma z' + \frac{2P}{\pi z'}\right)\sin\alpha = c_r + \left(\gamma z' + \frac{2P}{\pi z'}\right)\cos\alpha \tan\phi_r \quad (6.1)$$

where $\sigma = \left(\gamma z' + \frac{2P}{\pi z'}\right)\cos\alpha$: normal stress (kN/m²) (with $\sigma_p = \frac{2P}{\pi z'}\cos\alpha$ is the normal stress of overburden based on the Boussinesq equation (Murthy 2003).

$\tau = \left(\gamma z' + \frac{2P}{\pi z'}\right)\sin\alpha$ shear stress (kN/m²) (with $\tau_p = \frac{2P}{\pi z'}\sin\alpha$ is the shear stress of overburden based on the Boussinesq equation (Murthy 2003)).

γ   unit weight of soil (kN/m³)
z'   the projected thickness (with $z' = a\tan(\beta - \alpha)/\cos\beta$) (m).

Equation 6.1 could be rewritten as

$$\frac{c_r \pi z'}{\gamma \pi (z')^2 + 2P} = \sin\alpha - \cos\alpha \tan\phi_r \quad (6.2)$$

On the other hand. The factor of safety regarded to soil strength was defined by

$$\tan\phi_r = \frac{\tan\phi}{F_s} \text{ and } c_r = \frac{c}{F_s} \quad (6.3)$$

Substituting Eq. 6.3 into Eq. 6.2 we had

$$\frac{\frac{c}{F_s}\pi z'}{\gamma \pi (z')^2 + 2P} = \sin\alpha - \cos\alpha \frac{\tan\phi}{F_s} \quad (6.4)$$

Based on the equilibrium theory, the depth of occurrence of soil creep, where reach to critical equilibrium, was determined by a factor of safety equal to 1 and $z' = z'_{cr}$. We had

$$\gamma \pi (z')^2 \sin\alpha(\tan\phi - \tan\alpha) + c\pi z' \tan\alpha + 2P\sin\alpha(\tan\phi - \tan\alpha) = 0 \quad (6.5)$$

Equation 6.5 could be rewritten when z' is replace by $z' = a\tan(\beta - \alpha)/\cos\beta$, we have the equation that defines the creep plane by the length from the toe of slope to the considered point, a, and α as follows:

$$\gamma\pi a^2 \sin\alpha(\tan\phi - \tan\alpha)(\tan(\beta - \alpha))^2 + c\pi a \tan\alpha \tan(\beta - \alpha)\cos\beta$$
$$+2P\sin\alpha(\tan\phi - \tan\alpha)(\cos\beta)^2 = 0 \qquad (6.6)$$

On the other hand

On the other hand, if replace the driving shear stress component, $\tau_p = \frac{2P}{\pi z'}\sin\alpha$, by the critical stress level conducted from laboratory creep test, τ_r, the equation that allows estimating the critical slope angles that soil creep may take place could be achieved as below:

$$\tau_r \pi a \cos\beta \tan(\beta - \alpha) - \pi\gamma a^2(\tan(\beta - \alpha))^2 \sin\alpha - 2P(\cos\beta)^2 \sin\alpha = 0 \qquad (6.7)$$

6.3.3 Logarithmic Creep Modeling

The logarithmic creep models were developed based on results of triaxial creep tests and may be simply described by the following relation:

$$\tau_m = A\ln t + B \qquad (6.8)$$

with A, and B are logarithmic constants. They can be determined from experimental fitting as follows:

$$A = \frac{\sum \tau_{m_i} \times \ln t_i - (\sum \ln t_i \times \sum \tau_{m_i})/n}{\sum(\ln t_i)^2 - (\sum \ln t_i)^2/n}$$
$$B = \left(\sum \tau_{m_i} - A \sum \ln t_i\right)/n$$

and t is time variable measured in units such as second, minute, hours, or year.

6.4 Results and Discussions

6.4.1 Factors Inducing Soil Creep

Influence of soil engineering properties

The stress-strain plot of the soil sample obtained from the drained triaxial is shown in Fig. 6.6. Lower peak strength of soil sample with higher water content confirmed water content as an important factor that controlling the soil strength. Hence, soil

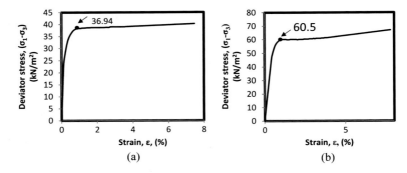

Fig. 6.6 Stress-strain curves of soil sample. **a** Soil with $w = 49\%$. **b** Soil with $w = 39\%$

strength decreases with increase of water content. On the other hand, decreases in shear strength of the soil can be expected if there are decreases in the effective stress and the cohesion of the soil.

The results of direct shear tests and evaluated friction angle and cohesion of the soil samples are shown in Fig. 6.7. Based on the results, volcanic residual soil in the research has low cohesion values (i.e., 1–19 kN/m^3, assumed to nearly zero when $\ll 100$ kN/m^2 (Potro and Hurlimann 2010)), and the internal friction angle in the range of 15.5–31.7°. Engineering properties of the soil samples were shown in Table 6.1. Based on the results, almost all the soil samples had the fine-grain fraction more than the coarse-grain fraction (i.e., fine content is greater than 50%). The significant fine content indicates the significant degree of weathering due to tropical climate. Moreover, high degree of weathering is also reflected by low density or high void ratio due to porous soil structure. According to (Culling 1963), the probability of creep displacement depends on the available pore space of the soil, which is reflected by the soil dry density. In the research area, the soils have moderate density of 1.69–1.82 g/cm^3 and relatively high void ratios of 0.92–1.1. The soil will, therefore, have higher susceptibility to creep.

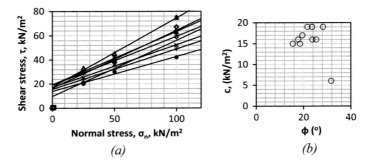

Fig. 6.7 a Results of direct shear tests. **b** Plot of cohesion and friction angle of the soil samples

Table 6.1 Engineering properties of the soil samples

Soil properties		
Sand	%	32.5–49.5
Silt	%	15.2–25.3
Clay	%	26.5–36.5
Water content (%)	w	30.9–49.43
Density (g/cm^3)	ρ	1.69–1.82
Dry density (g/m^3)	ρ_d	1.26–1.39
Void ratio	e	0.92–1.1
Porosity (%)	n	48.0–52.5
Liquid limit (%)	LL	41.9–48.2
Plastic limit (%)	PL	30.8–34.7
Cohesion (kN/m^2)	c	1–19
Friction angle (°)	ϕ	15.5–31.7
Drain peak strength (kN/m^2)	S_d	36.94–60.59
Young's modulus (MPa)	E	6.05–6.51
Poison's ratio	v	0.344–0.351

Additionally, the soil samples had liquid limit and plastic limit in the range of 41.9–48.2% and 30.8–34.7%, respectively. The soil was classified as medium swelling potential. Shrinking–swelling activities of the soil induced by moisture change can be one of the causes generating down-slope soil creeping. Therefore, the expansion of clay in the volcanic residual soil may be partly responsible for the occurrence of soil creeping in the research area.

Afterward, soil engineering properties including porosity, grain size, Atterberg limits are greatly dependent on the soil structure and specific interaction between mineral particles (Wesley 2009). Correspondingly, the mineralogical composition plays an important role that can explain the nature of soil engineering properties. In the other words, creep behavior is strongly influenced by soil engineering properties which are functions of mineralogical composition. Low strength parameters (i.e., cohesion, c, and friction angle, ϕ) of the soil samples compared with previous publications could be explained by the mineralogical composition and relatively high water content of soil samples, especially some samples have a water content greater than its liquid limit. Therefore, in the next section, the paper presents the result of soil mineralogical composition and its influence to soil creep behavior.

Influence of soil mineralogical composition

Figure 6.8 shows the X-ray diffraction pattern of the clay fraction of selected samples. Based on Fig. 6.8, air dried clay showed 10.4 Å, 10.3 Å, and 10.0 Å peaks of samples 1, 2, and 3, respectively. Ethylene glycol solvation resulted in swelling those peaks to 10.6 Å, 10.6 Å, and 11.1 Å, respectively. In additional, the 002 reflection of ethylene glycol treated clay changed from a relatively sharp peaks to broad peaks. It is clear

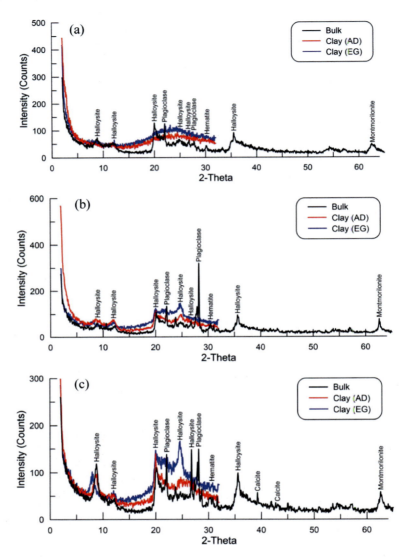

Fig. 6.8 The XRD pattern of the clay fraction of the soil. **a** Sample 1 at bore hole 1. **b** Sample 2 at bore hole 2. **c** Sample 3 at bore hole 3

indication that the ~10 Å reflection is due to the 001 reflection of hydrated halloysite. Accordingly to (Delvaux et al. 1990), the hydrated halloysite has the water-interlayer which reaches maximum swelling potential at ~10 Å. That means hydrated halloysite cannot be swell even through under ethylene glycol saturated. Such swelling peaks of ethylene glycol treated clay, therefore, indicated the presence of 2:1 clay which is high swelling potential. Accordingly, it is suggested that there must be the presence

Fig. 6.9 Interstratified haloysite-smectite layer (developed from Delvaux et al. (1990))

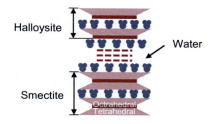

of a mixed-layer clay composed of a 10 Å hydrated halloysite and a 2:1 swelling clay mineral. On the other hand, the 015, 027, and 028 reflections (i.e., ~1.5 Å) of sample 1, 2, and 3, respectively, exhibit the presence of smectite (i.e., montmorillonite). The mixed-layer is, therefore, an interstratified halloysite-smectite layer (Fig. 6.9). This finding also agrees with conclusions of (Delvaux et al. 1990) after taking a series of XRD tests for hydrated halloysitic soil clays. He proposed that the presence of mixed-layer can be detected in the XRD patterns following these steps: 1, diffraction bands of thermal treated clay may be observed around 10 Å; 2, the ethylene glycol saturated clay show a differential swelling from the 10 Å refection and may reach to 13 Å; 3, persistence of broad diffraction peaks instead of sharp peaks.

Morphological characteristic of clayey samples has been presented by Scanning electron microscope (SEM) micrographs with magnification of ×20.000 and spatial resolution of 1 μm (Fig. 6.10). It is visible that the major morphology of soil samples is platy form and minor amount tubular and spheroidal forms. According to Lazaro (Lazaro 2015), halloysite crystals display variation in morphology depending on weathering conditions and geological origin. The most common structural form of halloysite is tubular and spheroidal those are preferred to be products from weathered volcanic ashes and pumice. Wada and Mizota (1982) explained the rolling mechanism of tubular halloysite as the dimension misfit between larger tetrahedral and smaller octahedral sheets. Moreover, Lazaro (2015) noticed iron content influences significantly the halloysite morphology. Generally, tubular halloysite content a low concentration of iron (i.e., 0–3%). As iron content increase, length of halloysite tube decreases relatively. Dixon and McKee (1974) considered tubular as characteristic of a dehydrated halloysite that expresses X-ray 001 peaks at ~7.14 Å. With high iron content halloysite (especially when Fe content >4%), substitution of a large amount of Fe^{3+} ion in the octahedral sheet would reduce the misfit between tetrahedral and octahedral sheets. As the result, the Fe-rich halloysite displays planar or platy form (Soma et al. 1992). Platy halloysite has been detected early by Quantin (1991) as weathering product in volcanic ash soils, volcanic pyroclastic when the weathering environment is rich in silica.

Results of energy-dispersive detector analysis (EDS) including elemental composition, compound composition and those quantity areas showed in Table 6.2. The EDS results reported the presence of Si, Al, and Fe in the soil samples with relative low Si/Al ratio ranging from 1.01 to 1.14, indicating that 1:1 clay mineral was the dominant portion. Moreover, the Fe content is relatively high (i.e., 6.39–6.66%). Delvaux

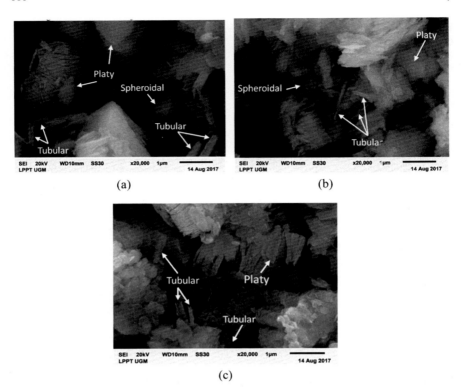

Fig. 6.10 SEM micrographs of the soil samples. **a** SP1 (at BH1, 1.5–2.0 m). **b** SP2 (at BH2, 1.5–2.0 m). **c** SP3 (1.5–2.0 m)

Table 6.2 The EDS analysis results of soil samples

	SP1	SP2	SP3	*Molar ratio	SP1	SP2	SP3
C	25.48	40.17	41.95	C	25.48	41.95	41.95
O	35.17	27.84	26.99	Al_2O_3	28	22.27	22.27
Al	14.82	12.49	11.79	SiO_2	37.59	27.56	27.56
Si	17.57	13.07	12.88	K_2O	0.37	–	–
K	0.31	–	–	FeO	8.56	8.22	8.22
Fe	6.66	6.42	6.39	SiO_2/Al_2O_3*	2.28	2.01	2.1
C	25.48	40.17	41.95				
Si/Al*	1.14	1.01	1.05				

6 Slope Creep Instability in Krajang Lor Village ... 111

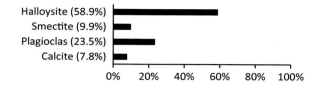

Fig. 6.11 Mineralogical composition of the soil samples

et al. (1990) noticed that high Fe concentration (i.e., %Fe > 5) presents in high-charge halloysite or hydrated halloysite. This determination confirms the results of XRD analysis that there is the presence of hydrate halloysite clay.

Figure 6.11 shows the mineralogical composition of soil samples. The clay fraction is dominated by hydrated halloysite (Fig. 6.11) (i.e., 58.9%) with a minor amount of smectite (montmorillonite) (i.e., 9.9%). Hydrated halloysite is less stable than tubular dehydrated halloysite. The interlayer water of platy hydrated halloysite is weakly held and, thus, can be easily irreversibly lost under effects of stress or heating. The clay particle is, then, rearranged to reach the more stable form. This process partly resulted in soil creep in the research area. On the other hand, although halloysite was the main clay component in the soil, an amount of smectite of less than 10% could give an extremely creep behavior. The reason is smectite has the highest specific surface area of approximately 800 m^2/g while that of hydrated halloysite is only approximately 65–70 m^2/g (Tuladhar et al. 2004). In the other words, the activity of smectite is approximately 12 greater than the hydrated halloysite. The soil containing only 10% of smectite has, therefore, creep behavior 3 times greater than that with purely hydrated halloysite. It indicates the important role of smectite in the research volcanic residual soil.

Influence of loads

Initial condition of FEM-slope stability analysis

In order to understand the influence of loadings to existing creep slope instability in the research area. slope stability analyses were conducted using FEM (i.e., Plaxis 8.2 software) for both scenarios with long-term conditions of loading and without loading. Based on direction of the crack distribution, two cross sections, so call section AA' and section BB' which have different slope inclinations, were chosen for numerical analysis as illustrated in Fig. 6.12. Figure 6.13 illustrates the geometry and the two-dimensional finite element meshes composed from 303 15-noded elements used for example of BB' cross section. The boundary condition and soil parameters are also introduced in Fig. 6.13.

Graphical outputs presenting shading of the horizontal displacements and localization potential slip surface of both numerical analyses those with loading and without loading are shown in Figs. 6.14 and 6.15, respectively.

In case of vertical loading of 5kN/m^2 acting all along the slope, Fig. 6.14 indicates the displacements of soil particle concentrating at the top and middle of the slope. Those two instable zones locate at the elevations of 390–365 m and 360–345 m

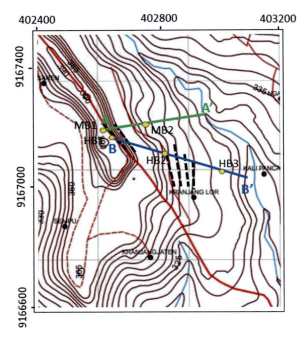

Fig. 6.12 Chosen research cross sections for numerical analyses

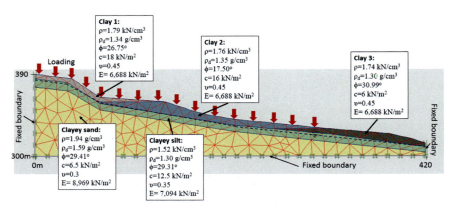

Fig. 6.13 Initial condition of FEM models in case of with loading (example for the BB′ cross section)

with the circular form of slip surfaces. In AA′ cross section, depths of potential slip failure at the top and middle of the slope are approximately 11.526 m and 2.737 m, respectively, whereas in BB′ cross section those potential slip surfaces appear at depth of 7.353 m and 2.500 m, respectively. The extreme displacement of AA′ cross section is greater than that of BB′ cross section (i.e., 234.1 m and 185.9 m, respectively).

6 Slope Creep Instability in Krajang Lor Village ...

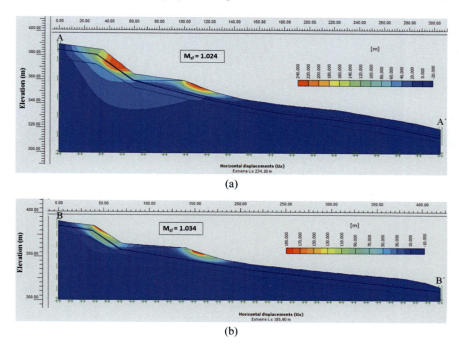

Fig. 6.14 Shading displacements of FEM-slope stability analyses with loading. **a** AA' cross section. **b** BB' cross section

The different extreme displacement is due to the different geometry as the AA' cross section has higher slope inclination. However, the factor of safety of both cross sections is almost similar and very close to 1. Low values of factor of safety indicate the slopes are in the state of impending failures.

In case of numerical analysis without loading as shown in Fig. 6.15. The displacements of soil particle concentrate only on the top of the slope but are absent on the middle of the slope. The factor of safety in AA' and BB' cross sections are 1.563 and 1.628, respectively. Those values of factor of safety are comparatively high with the case of numerical analysis with loading (i.e., approximately 1.0). It indicates a significant influence of loading on stability of the slope in Krajang Lor Village. On the top of the slope which has high slope inclination, the loads have been causing decreasing factor of safety as well as increasing thickness of potential movement zone. At the middle of the slope which has gentle inclination, the load is responsible for the occurrence of new unstable zone. Based on those results, the loads have been considered as the main factor that induces slope creeping in the Krajang Lor Village.

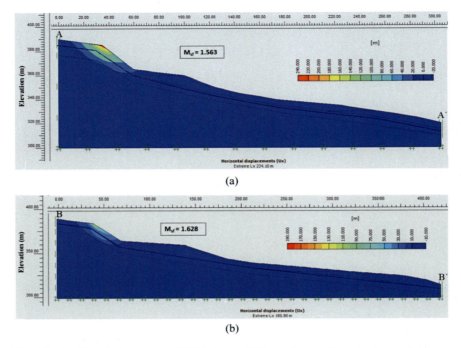

Fig. 6.15 Shading displacements of FEM-slope stability analyses without loading. **a** AA′ cross section. **b** BB′ cross section

6.4.2 The Thickness of Creep Zone and Critical Slope Angles

The thickness of creep zone

In order to evaluate the thickness of creep zone, a slope stability model based on Mohr-Coulomb failure criterion that considers stress distribution as factor inducing creep was developed as expresses in Eq. 6.6, which was the relation between the engineering properties of soil (i.e., c, ϕ, γ) and the slope parameters (i.e., β, α, a). Input for numerical calculation includes soil properties that were the same as slope model used from FEM-slope stability analysis (Fig. 6.13). The loops computation was conducted for both AA′ and BB′ cross section (Fig. 6.12) and the calculated depth of creep zone on AA′ and BB′ cross sections are shown in Figs. 6.16 and 6.17, respectively.

Based on the results, there are two active creep zones located beneath soil surface at the elevations of 390–365 m (so-called active creep zone I, ACZ I) and 360–345 m (so-called active creep zone II, ACZ II), corresponding to locations of local main road and housing area of Krajang Lor Village, respectively. ACZ I appears on AA′ and BB′ cross sections with maximum thickness of 10.368 m and 6.537 m, respectively. ACZ II appears on AA′ and BB′ cross sections with maximum thickness of 2.4 m and 2.271 m, respectively.

6 Slope Creep Instability in Krajang Lor Village …

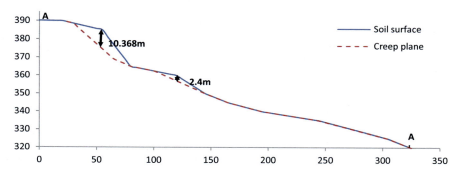

Fig. 6.16 Evaluated depth of creep zone on cross section AA′

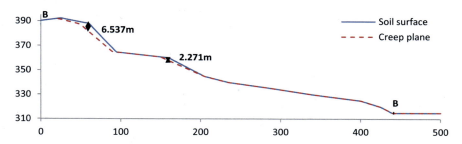

Fig. 6.17 Evaluated depth of creep zone on cross section BB′

The comparison between evaluated thickness of creep zone using proposed model and calculated depth of slip surface from FEM analysis conducted in sup-Sect. 4.1 are presented in Table 6.3. The degrees of differential are in a range of 9.16–21.31% indicating that the thickness of creep zone estimated from the proposed model coincides closely with estimated slip surface from FEM-slope analysis. The proposed stability model based on Mohr-Coulomb failure criterion can be applied to estimate the potential creep zone. The advantage of the proposed model in comparison with the FEM analysis is allowing estimating the thickness of creep zone without assumption of circular slip surface as well as need less input parameters (i.e., only ϕ, c, γ,

Table 6.3 Depth of creep zone

Location		FEM analysis	Proposed model	Degree of differential (%)
AA′ cross section	Top of the slope	11.526	10.368	10.05
	Middle of the slope	2.737	2.400	12.31
BB′ cross section	Top of the slope	7.353	6.537	11.10
	Middle of the slope	2.500	2.271	9.16

P and geometry parameters) while FEM analysis needs many engineering parameters those must be conducted in costly and timing tests. However, the proposed model remains some disadvantages due to unavailability for analysis under dynamic conditions such as changing of soil properties and groundwater due to season.

Critical slope angles

The critical slope angle is identified as the maximum slope inclination in which the soil slope undergoing creep (i.e., considered as primary and secondary creep stages) but without slope failure (i.e., considered as tertiary creep stage) under existing climate and land uses. At critical slope angle, the slope would occur slip the surface and begin to slide down. This critical slope angle is quite variable depending on soil materials (Skempton and Delory 1952). That makes difficulty in researching about the critical slope angle for a particular soil type. On the other hand, it has been a lack of research on laboratory testing methods for evaluating values of the critical slope angle. Therefore, this paper has been tried to develop a method to evaluate the critical slope angle that the Krajang Lor soil can support. However, because of limitation on testing of critical slope angle, the proposed method remains invalidated. But it may provide a better understanding of soil behavior by comparing the results of prediction with the existing slope angle in the research area.

Numerical calculation of potential critical slope angles was conducted based on Eq. 6.7. Input for numerical calculation includes soil properties that were the same as slope model used from FEM-slope stability analysis (Fig. 6.13). Figure 6.18 shows the results of numerical calculation of slope at elevation of 300–330 m, 330–360 m and 360–390 m, respectively. Based on the results, the soil near the toe of the slope can support slope inclination reaching to 41.7°; while those at the top and the middle slope can support slope inclination reaching to 38.5° and 23.9°, respectively. The soil at the middle slope can support for slope with low critical slope angle (i.e., 23.9°) indicates a weakened zone developed in housing area of the Krajang Lor Village where the soil has very weak properties. The comparison between the predicted critical slope angle and existing slope angle in Krajang Lor Village is shown in Table 6.4. The results show that existing slope angles are less than the critical slope angles needed for slope failure. However, different values between existing slope angle and the predicted critical slope angle at the top and middle slop are relatively low ranging from 4.7° to 8.6°. This suggests that any change in slope inclination due to cuttings and land used may lead the soil to the critical stage, then, initiating slope failure.

6.4.3 Developmental Creep Displacement

Step 1: Identifying the distribution of stress level with depth

Total stress at a certain depth in a soil mass is generated considerably by combination of surface loads and overburden. In a slope, this stress separates into 2 components which are: normal stress (i.e., the component perpendicular with soil surface) and

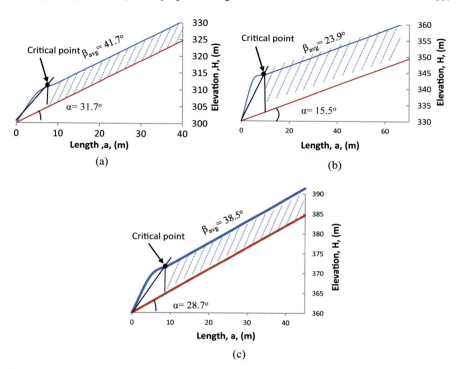

Fig. 6.18 Numerical calculation of critical slope angle. **a** The toe (elevation of 300–330 m), **b** the middle slope (elevation of 330–360 m). **c** The top slope (elevation of 360–390 m)

Table 6.4 Comparison between evaluated critical slope angles and existing slope angles in Krajang Lor Village

Position	Elevation (m)	Evaluated critical slope angle (°)	Existing slope angle (°)	
			AA′ cross section	BB′ cross section
The top slope	360–390	38.5	33.8	29.7
The middle slope	330–330	23.9	18.4	15.3
The toe	300–330	41.7	13.7	11.3

shear stress (i.e., the component parallel with soil surface, down-slope direction). Stress level generated in soil mass is estimated by the ratio between driving shear stress and the soil shear strength as

$$\frac{\tau}{\tau_f} = \frac{\left(\gamma z' + \frac{2P}{\pi z'}\right)\sin\alpha}{c + \sigma\tan\phi} \quad (6.9)$$

where

τ_f	is the yield strength of soil (kN/m^2)
τ	is the driving shear stress
σ	is the normal stress (kN/m^2)
γ	unit weight of soil (kN/m^3)
c	is the cohesion of soil
ϕ	is the friction angle of soil
tanϕ	is the coefficient of internal friction
P	is the stress of surface loads per unit length
z'	is the soil depth.

The soil in Krajan Lor Village is subjected under surface loads of approximately 5 kN/m^2. Subjecting $\gamma \sim 16.9$–1.82, $c \sim 0.06$–0.19, $\phi \sim 15.5$–31.7 into the Eq. 6.9, the distribution of stress level with soil depth is obtained as in Fig. 6.19. Based on the results, the stress level is the highest (i.e., 81% of the soil shear strength) at the soil surface and decreases to 60% of the soil shear strength at 19 m depth.

Step 2: Development of logarithmic creep model

The logarithmic creep model was developed based on the results of triaxial creep tests. Experimental curves and logarithmic fitting are shown in Fig. 6.20. As the results, the computational formulations of stress levels of 80% and 60% of the soil shear strength, which are $\varepsilon = 0.088\ln t + 4.398$ and $\varepsilon = 0.028\ln t + 3.535$, respectively.

Based on the results of creep tests, the strain-time curve divides into three stages (i.e., primary, secondary, and tertiary). Due to limited creep testing duration, creep observation had only the primary and secondary creep stages and the absence of creep rupture could be achieved at deviatoric stress level of 20, 40, and 60%. At stress level of 80%, there was evidence of tertiary creep at the end of testing duration (i.e., after 166 h). And at stress level of 90%, the soil specimen failed after 20 h.

On the other hand, the critical stress level of soil creep could be established. The critical stress lever could be considered as that cause creep rupture during testing duration. Correspondingly, the critical stress levels of haloysite-rich soil collected in the Krajang Lor Village are 80%. Currently, loadings from housing and human activities in the Krajang Lor Village are about 5 kN/m^2 producing a stress level of approximately 14% of the soil shear strength. Based on the result, if there is increase

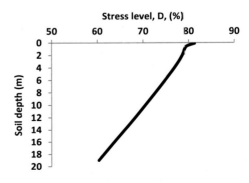

Fig. 6.19 Distribution of stress level with soil depth in Krajang Lor Village

6 Slope Creep Instability in Krajang Lor Village … 119

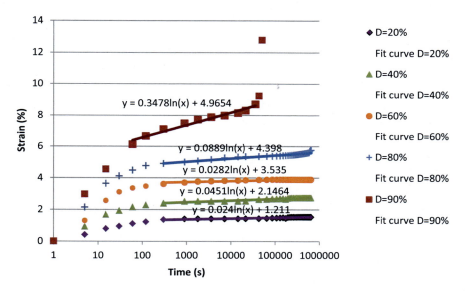

Fig. 6.20 Logarithmic fitting curves of halloysite-rich soil ($\tau = 60.54$ kN/m^2; $w = 39\%$; $\rho_d = 1.29$; $\sigma_3 = 40$ kN/m^2)

in housing or loads, for example, up to 3 times, the creep displacement could increase 30%. Once loadings are increased 6 times, the slope may lead to a rupture.

Step 3: Prediction of creep displacement

Estimation of creep displacement of soil mass in the research area is shown in Fig. 6.21. Creep displacement has been calculated for up to the process of 60 years since first creep initiated. Creep displacements of the first 7 days are based on results

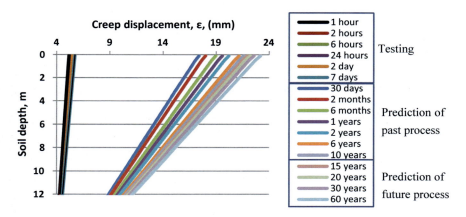

Fig. 6.21 Creep displacement of soil mass in Krajang Lor Village

of creep testing while those of the rest duration is prediction. There are two stages of creep displacement prediction.

Prediction of pass process: Assuming creep has been initiated since 2006 as reported by local government, the existing creep has been going for 10 years. In this stage, estimation of existing creep displacement is conducted. Based on the result, at the moment (i.e., 10 years since creep soil creep initiated) creep displacement at the soil surface are 21.6 mm; and those at the creep plane (i.e., 10.5 m depth) are 11.9 mm and 0.06 mm/year, respectively.

Prediction of future process: Development of creep in the future process is also estimated. In this paper, the author gives predicting examples of 15, 20, 30, and 60 years after soil creep initiated (or 5, 10, 20, and 50 years from the moment of this research). Based on the result, creep displacement at the soil surface will increase to approximately 22 mm and approximately 23 mm after 15 years and 60 years, respectively.

The results of prediction are verified by field investigation of tension cracks observed from preliminary field investigation. Accordingly, the major crack on pavement of the local road has approximately 21 mm width (Fig. 6.3). A good agreement between the prediction results and field measurement indicates the results of prediction are acceptable.

6.5 Conclusions

The paper reported the problem of slope creep instability in the Krajan Lor Village, Salaman District, Magelang Regency, Central Java, Indonesia. In this paper, field investigation and laboratory testing were carried out in order to explain the cause of soil creep while slope stability modeling based on Mohr-Coulomb failure criterion and logarithmic creep modeling were conducted to predict the development of soil creep displacement. Major findings of the paper are summarized as follows:

1. Main factors causing soil creep included poor engineering properties of the soil (i.e., high water content, low dry density, and low shear strength parameters), irreversibly dehydration of hydrated halloysite clay as well as swell-shrink activities of smectite clay and loadings of about 5 kN/m^2 as a primary trigger of creep initiation. When the load in this research area increases up to 3 times the current load, soil strain due to creep increase up to 30%. When the load in this research area increase up to 6 times the current load, the slope will fail eventually. A good plan of local land uses management are, therefore, required promptly.
2. The thickness of creep zone in the research area was 10.368 m and critical slope angles were 41.7° at the toe of the slope; 23.9° at the middle slope and 38.5° at the top of the slope.

3. Soil creep displacement at the soil surface was predicted increasing to approximately 22 mm after 15 years and 23 mm after 60 years.

References

Culling WEH (1963) Soil creep and the development of hillside slopes. J Geol 71:127–161

Delvaux B, Herbillon AJ, Vielvoye L, Mestdagh MM (1990) Surface properties and clay mineralogy of hydrated halloysitic soil clays. II: evidence for the presence of halloysite/smectite (H/Sm) mixed-layer clays. Clay Mineral 25:141–160

Dixon JB, McKee TR (1974) Internal and external morphology of tubular and spheroidal halloysite particles. Clays Clay Miner 22:127–137

Fleming RW, Johnson AM (1975) Rate of seasonal creep of silty clay soil. Q J Eng Geol 8:1–29

Lazaro BB (2015) Halloysite and Kaolinite: two clay minerals with geological and technological importance. Rev Real Acafemia De Ciencias Zavagoza 70:7–35

Lytton RL, Dyke D, Mathewson CC (1980) Creep damage to structures on expansive clay slopes. A Report from the Texas A&M Research Foundation, Department of Civil Engineering, Texas A&M University

Murthy VNS (2003) Chapter 6: stress distribution in soils due to surface loads. In: Geotechnical Engineering—Principles and practices of soil mechanics and foundation engineering. Marcel Dekker, Inc., pp 173–203

Owens IF (1967) Mass movement in the Chilton valley. Thesis (Master). Arts in Geography

Potro RD, Hurlimann M (2010) The origin and geotechnical properties of volcanic soils and their role in developing flank and sectors collapse. In: Volcanic rock mechanics: rock mechanics and geo-engineering in volcanic environments. Taylor and Francis Group, pp 159–166

Quantin P (1991) Specificity of the halloysite-rich tropical or subtropical soil. O.R.S.T.O.M. Fonas Document (31994):16–21

Rahardjo W, Sukandarrumidi, Rosidi HMD (1995) Geological map of the Yogyakarta sheets, Jawa. Geological Research and Development Center. Geological Survey of Indonesia Mistry of Mines

Robert WD (2006) Foundation engineering handbook: design and construction with the 2006 international building code. American Geotechnical San Die-go, California, Copyright The McGraw-Hill Companies, Inc. 2006 under li-cense agreement with Books24 × 7

Skempton AW, Delory FA (1952) Stability of natural slopes in London clay. In: Proceedings of the 4th international conference on soil mechanic, London, pp 378–381

Soma M, Churchman GJ, Theng BKG (1992) X-ray photoelectron spec-troscopic analysis of halloysites with different composition and particle morphology. Clay Miner 27:413–421

Swanston DN (1974) Chapter 5: soil mass movement. In: The forest ecosystem of southeast Alaska, USDA forest service general technical report PNW-17

Tuladhar GR, Marui H, Tiwari B (2004) Long term stability of the land-slides along mudstone dominated area having NaCl type pore water. In: Ehrlich, Fontoura, Sayao (eds) Landslide: evaluation and stabilization. Taylor & Francis Group, London, pp 649–654. ISBN 04-1535-665-2

Wada SI, Mizota C (1982) Iron-rich halloysite (10 Å) with crumpled lamellar morphology from Hokkaido. Jpn Clay Clay Miner 30(4):315–317

Wesley L (2009) Behavior and geotechnical properties of residual soils and allo-phane clays. Obras y Proyectos 6:5–10

Ziemer RR (1977) Measurement of soil creep by inclinometer. In: Engineering Technical Report, Forest Service, U.S, vol 20013. Department of Agriculture Washington, D.C., pp 1–10

Chapter 7
Application and Feedback Analysis of the Freeway Slope Maintenance Management System in Taiwan

San-Shyan Lin, Wen-I Wu, Tsai-Ming Yu, Chia-Yun Wei, Lee-Ping Shi, and Jen-Cheng Liao

7.1 Introduction

There are 1049.7 km long freeways in Taiwan. Along the freeway, there are 2,567 slopes that include 927 cut slopes, where 148 slopes were installed with inclinometers, load cells or other monitoring gages and 160 slopes were stabilized with 30,608 ground anchors.

On a sunny afternoon on April 25, 2010, the dip slope mainly stabilized by ground anchors at 3.1 k of the number 3 freeway failed with a large amount of slope debris covered on all the freeway lanes. The covered area was approximately 200 × 60 m^2 with a total debris volume estimated at about 100,000 m^3. One cross over the bridge also collapsed resulting from the thrust of the debris.

After the incident of the 3.1 k slope failure, the Taiwan Freeway Bureau adopts a life-cycle based maintenance and management system to ensure the slopes along the freeway stays stable and adapts to different environmental conditions. It expects to ensure safety for the freeway users at all times.

In the following, the maintenance and management system is briefed and then explained in detail by a case example.

S.-S. Lin (✉)
National Taiwan Ocean University, Keelung, Taiwan
e-mail: sslin46@gmail.com

W.-I. Wu · T.-M. Yu · C.-Y. Wei
Taiwan Area National Freeway Bureau, New Taipei City, Taiwan

L.-P. Shi · J.-C. Liao
Taiwan Construction Research Institute, New Taipei City, Taiwan

© The Author(s), under exclusive license to Springer Nature Singapore Pte Ltd. 2023
H. Hazarika et al. (eds.), *Sustainable Geo-Technologies for Climate Change Adaptation*, Springer Transactions in Civil and Environmental Engineering,
https://doi.org/10.1007/978-981-19-4074-3_7

7.2 Slope Maintenance

The slope maintenance manual of the Freeway Bureau in Taiwan (Freeway Bureau 2017) provides a guideline and SOP for the engineer when dealing with slope related cases. It contains several chapters that include slope inspection and monitoring, anchor detection, slope grading; slope maintenance and restoration; personnel management and education; slope management meeting; establishment and application of a slope maintenance information management system. The manual is updated each year under a supervision meeting after the suggestions collected from the engineers.

In order to have an efficient and economical management, the slopes are classified into four different categories, that are grades A, B, C and D. The categorizations are based on the slope site inspection, monitoring measured data, anchor detected data and stability assessment results. The four categories of slope are:

Grade A: The slope appears to have obvious signs of instability to eyes. It needs to take some effective stabilization actions along with detailed inspection and monitoring.

Grade B: The slope is possibly unstable that can be observed by the eyes. Some maintenance, reinforcement and curing may need to take along with increasing inspection and monitoring.

Grade C: The slope has no obvious signs of instability. It needs some inspections, regular maintenance and monitoring if needed.

Grade D: The slope is stable. It only needs to take casual inspections.

Along the freeway all of the slopes were inspected and evaluated after the incident of the 3.1 k case, the slopes classified as grades A and B were repaired, reinforced or installed with the monitoring system. The frequency on slope perambulation, monitoring and anchor inspection all depends on the grade of the slope. Take the slope perambulation as an example, the inspection frequency is at least once a month, once a season, once a year and once in three years for the respective slopes of grades A, B, C and D. The grade of each slope is adjustable after investigation, monitoring data analysis, anchor detection data analysis and/or reinforcement. Detailed classification procedures are given in Fig. 7.1. Any collected data of each slope is stored in the cloud. In addition, any observation from the site inspection can be uploaded through the internet at the site via a pad.

Figure 7.2 shows the statistical information after slope inspection in northern Taiwan in 2014 and 2017. It is seen that the percentage of drainage shortage increased from 36.8 to 50.3%. On the other hand, the percentage of stability facility shortage decreased from 26.5 to 15.9%. The drainage inspection becomes critical work for the stability of a slope during the rainy or monsoon season.

For the anchor detection, it includes function detection such as test location selection, anchor head components inspection, endoscope check for tendon corrosion and lift-off test. An anchor grade evaluation is required after the detection (Freeway Bureau 2017). The anchor detection results between 2011 and 2013 are shown in

7 Application and Feedback Analysis ...

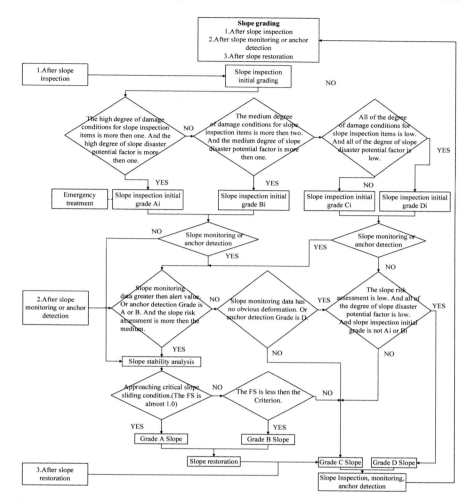

Fig. 7.1 Slope classification procedure

Fig. 7.3. It shows that 57–86% of the inspected anchors are in their working conditions. To prolong the serviceability of the ground anchor after detection, the anchor free end was grouted, the anchor head was covered with a galvanized plate and its internal space was filled with special grease.

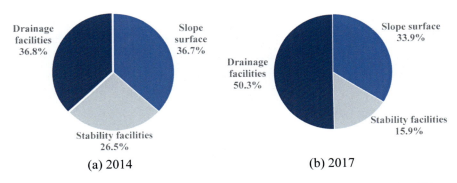

Fig. 7.2 Shortage statistics of main items of slope inspections in 2014 and 2017 at northern Taiwan

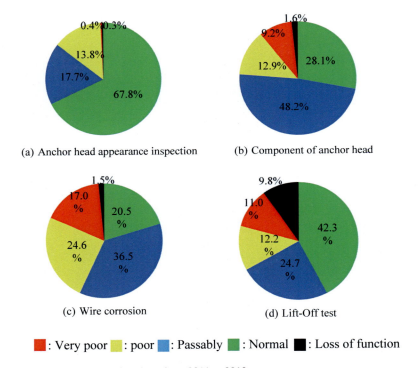

Fig. 7.3 Detection results of anchors from 2011 to 2013

7.3 Management System

The slope maintenance and management system (SMMS) includes two sub-systems that are the slope inspection operation system (SIOS) and the slope information sharing platform (SISP).

The SMMS system provides the functions of data collection, storing, classification, inquiry and the functions of situation reciprocation and processing tracing, waiting list notice and tracing, abnormal event notice, focus section monitoring and controlling etc. In addition, the system also increases the values of information which includes statistical analysis of inspection and maintenance records, accumulations of damage histories, slope grading and safety evaluation, etc.

The SIOS uses the pad as the operation tool during the inspection. It can (1) download the recent fundamental slope information and off-line map; (2) show the shortage of the location and the photo taken in the last inspection and (3) locate the shortage location with a picture. The shortage records that contain pictures can be sent back to SMMS. It can also record the track of inspection. The SISP is a large amount of document database which includes slope fundamental data, inspection data, maintenance data, detection and monitoring data, safety analysis and reinforcement data, disaster cases, education and training data, specification, manual and the others.

The system contains many integrated functions that are also used for developing part of the freeway disaster potential map. The system also includes the data of the accumulated rainfall and the thresholds established by the National Science and Technology Center for Disaster Reduction (NCDR) are used as an early warning reference value (Lin et al. 2017).

7.4 A Case Example

The studied case was a slope with mudstone and was originally classified as grade C. During a routine inspection trip in 2018, the deformation of the slope showed a bulging at the second stage of the slope was observed as shown in Fig. 7.4. By the inclinometer observed information, the shallow slide at depth 2.0 m below the slope surface was observed as shown in Fig. 7.5, at which the slope was re-graded as grade B. The authority did not take immediate action to solve the problem but enforce inclinometer monitoring frequency.

Three months later the slope failed after a period of heavy rainfall as shown in Fig. 7.6. After investigation, it was found the possible reason for the deformation was due to erosion of the weathered mudstone and seepage resulted from heavy rainfall. The slope became a case of grade A that required an emergent engineering reinforcement or repair.

The slope contains mudstone that is easy to be eroded away and disintegrated. Hence, in addition, to repair the failed area, the neighbor area of the slope was also reinforced by a reinforced concrete grid with additional 5 m long earth anchor. At the toe of the slope, it was installed with a 3 m retaining wall and 30 cm micro-piles as shown in Fig. 7.7. A finite element numerical analysis was carried out shown in Fig. 7.8, the estimated safety factor is higher than 1.9 even under heavier rainfall than the failed record. The slope after repairing is shown in Fig. 7.9 that is classified as grade C slope after repairing.

Fig. 7.4 A bulge second stage was observed

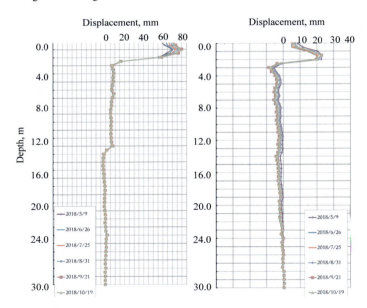

Fig. 7.5 The inclinometer monitoring data

7 Application and Feedback Analysis … 129

Fig. 7.6 Failed slope

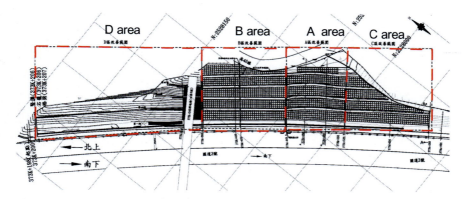

Fig. 7.7 Slope repair areas

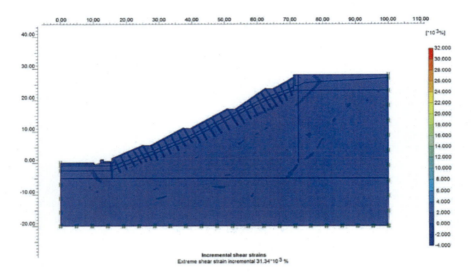

Fig. 7.8 Numerical analysis

Fig. 7.9 Slope after repairing

7.5 Conclusions

This paper introduced the maintenance and management system for the slopes along the freeway in Taiwan. The basic geographic and geological data, site inspection information and repairing record etc. for each slope is integrated into one system. The engineer can easily access the information from the system by a pad. A case example was used to demonstrate the usefulness of the system. It is believed that a serious slope failure such as the case at 3.1 k will not happen again along the freeway in Taiwan.

References

Freeway Bureau (2017) MOTC. Maintenance manual of freeway, Chapter 3

Lin BS, Wei JY, Shi LP, Liao JC, Lin XH (2017) Installation plans of rainfall stations for the maintenance management of freeway slopes. In: Proceedings of the 5th international conference on geotechnical engineering for disaster mitigation and rehabilitation. CTGS, Taipei, Taiwan, pp 385–390

Part II
Characterization of Geo-Materials

Chapter 8
Chemical and Mechanical Properties of Geopolymers Made of Industrial By-Products Such as Fly Ash, Steel Slags and Garbage Melting Furnace Slags

Tatsuya Koumoto

8.1 Introduction

More than 60 billion kgs of industrial by-products such as fly ash, steelmaking slags, and garbage melting furnace slags are generated every year in Japan (Koumoto 2019). Almost all of them seem to have been effectively used in mixtures with cement according to their chemical and mechanical properties in the field of civil work. Fly ash and slags can be used to create geopolymers in a process that emits less carbon dioxide than in the cement making process. This reduction in CO_2 emission is important because CO_2 is one of the substances known to contribute to global warming. In the future, further uses of these fly ash and slags must be explored.

The chemical mechanism for hardening composite materials by mixing aluminosilicate binders, such as fly ash, and slags with alkaline activators, such as liquid NaOH and sodium silicate, is known as a geopolymer reaction. The hardened composite material is called geopolymer (Davidovitz 1991). Geopolymers are produced by mixing two components of solid binders like fly ash, slags, etc. and liquid activators like NaOH, KOH, sodium silicate and so on (Buchwald 2006). Geopolymers have recently been developed to be used as a replacement for Portland cement concrete. The development of high compressive strength geopolymer using fly ash and slags will strongly contribute to the fields of construction, geotechnical engineering, and architecture.

This paper describes the chemical and mechanical characteristics of geopolymers by mixing fly ash and slags as binders with an alkaline solution of NaOH and sodium silicate as activators.

T. Koumoto (✉)
Saga University, Saga 840-8502, Japan
e-mail: koumotot@cc.saga-u.ac.jp

8.2 Materials

8.2.1 Preparation of Geopolymer Materials

The finer the particle size, the greater the compressive strength of geopolymers (Koumoto 2019).

In the present tests, all slags were ground after being air dried to a maximum particle size of 0.106 mm for effective chemical reaction.

8.2.2 Chemical Compositions of Geopolymer Materials

The main chemical compositions of the geopolymer materials are listed in Table 8.1. The seven kinds of geopolymer materials (two each of Fly ash, Slag1 (steel factory slags), Slag2 (garbage melting furnace slags) and one Acidproof cement) were the starting geopolymer materials, and seventeen mixture materials (two Fly ash and Fly ash, one Acidproof cement and Slag1, five Fly ash and Slag1, two Fly ash and Slag2, four Slag1 and Slag1 and three Slag1 + Slag2) were prepared as materials with a wide range of chemical compositions.

A triangular coordinate display for geopolymer materials listed in Table 8.1 is shown in Fig. 8.1. In Fig. 8.1 triangle is drawn for CaO, SiO_2 and Al_2O_3 + others. This display method helps us to understand the situation of chemical compositions of materials as binders visually. All materials listed in Table 8.1 are plotted inside the parallelogram.

8.2.3 Making Geopolymer Samples

Although several kinds of liquid sodium hydroxide and sodium silicate are available as activators, for the purpose of this research 48% NaOH (18 mol/L) and sodium silicate ($Na_2 \cdot nSiO_2$, $n = 3.2$) were chosen due to their commercial availability and high concentration, with the expectation to yield high compressive strength geopolymers. The compressive strength of geopolymers, q_u, is generally considered to be a function of the weight ratio of the activator to the binder, w, and the weight ratio of NaOH to sodium silicate, η. According to Koumoto (Koumoto 2019), the w_{opt} which is the optimum value of w yielding the ultimate compressive strength q_{umax} becomes a constant value of 0.4. In this research, geopolymer samples were made for η(0.0–1.0) at a constant value of $w = 0.4$. After mixing the activator and binder, the geopolymer samples in soft conditions were placed in plastic molds of diameter $D = 50$ mm and height $H = 100$ mm and vibrated slightly.

The geopolymer samples were removed from the molds after 1 or 2 days and cured at room temperature under dry conditions for 28 days.

8 Chemical and Mechanical Properties of Geopolymers Made …

Table 8.1 Chemical compositions of tested geopolymer materials

Geopolymer materials	Geopolymer samples	Chemical composition (%)					
		SiO_2	Al_2O_3	CaO	Fe_2O_3	MgO	SO_3
Fly ash	Reihoku	55.0	21.1	9.1	5.3	1.1	0.9
	Karita	38.8	24.3	19.5	1.6	0.5	6.6
Slag1	Koro	34.6	14.8	42.7	0.4	5.7	0.0
	Stainless	26.7	5.3	48.2	1.0	5.5	0.4
Slag2	Kazusa	34.2	13.2	42.0	2.6	1.9	0.7
	Narashino	34.2	13.9	39.3	3.7	1.8	0.6
Acidproof cement	Selament	49.8	13.6	26.7	2.6	*	*
A. cement + Slag1	SelaKoro	42.2	14.2	34.7	1.5	*	*
Fly ash + Fly ash	KariRei	46.9	22.7	14.3	3.5	0.8	3.8
	KariRei2	49.6	22.2	12.6	4.1	0.9	2.8
	ReiKoro	44.8	18.0	25.9	2.9	3.4	0.5
	ReiKoro2	41.4	16.9	31.5	2.0	4.2	0.3
Fly ash + Slag1	KariKoro	36.7	19.5	31.1	1.0	3.1	3.3
	KariKoro2	36.0	18.0	35.0	0.8	4.0	2.2
	KariKoro7	35.1	16.0	39.8	0.6	5.1	0.8
Fly ash + Slag2	ReiKazu	44.6	17.2	25.6	4.0	1.5	0.8
	ReiNara	44.6	17.5	25.6	4.5	1.4	0.8
	KoroSta5	28.0	6.9	47.3	0.9	5.5	0.3
Slag1 + Slag1	KoroSta2	29.3	8.5	46.4	0.8	5.6	0.3
	StaKoro	30.7	10.1	45.5	0.7	5.6	0.2
	StaKoro2	32.0	11.6	44.5	0.6	5.6	0.1
	KazuKoro	34.4	14.0	42.4	1.5	3.8	0.4
Slag1 + Slag2	KazuKoro2	34.5	14.3	42.5	1.1	4.4	0.2
	KazuKoro5	34.5	14.5	42.6	0.8	5.1	0.1

Note Koro = Ground blast furnace slag, Stainless: Ground stainless steel making slag, Kazusa and Narashlno = Garbage melting furnace slag
Mixture ratio of geopolymer samples
KariRei: Karita: Reihoku = 1: 1 in weight; KariRei2: Karita: Reihoku = 1: 2 in weight
ReiKoro: Reihoku: Koro = 1: 1 in weight; ReiKoro2: Reihoku: Koro = 1: 2 in weight
KariKoro: Karita: Koro = 1: 1 in weight; KariKoro2: Karita: Koro = 1: 2 in weight
ReiKazu: Reihoku: Kazusa = 1: 1 in weight; ReiNara: Reihoku: Narashino = 1: 1 in weight
KazuKoro: Kazusa: Koro = 1: 1 in weight; KazuKoro2: Kazusa: Koro = 1: 2 in weight
KoroSta2 = Koro: Stainlesss = 1: 2 in weight; KoroSta5 = Koro: Stainless = 1: 5 in weight

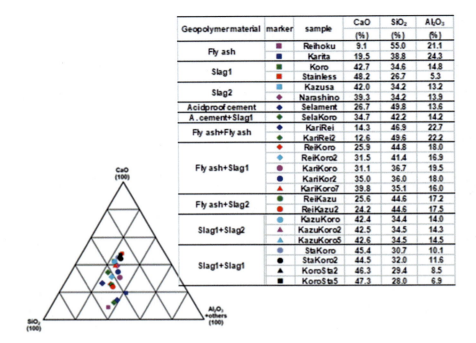

Fig. 8.1 Triangular coordinate display for geopolymer materials listed in Table 8.1

8.3 Tests and Results

8.3.1 Compression Tests

Before compression tests, physical properties of geopolymer samples such as diameter d, height h and weight W were measured to obtain characteristics of shrinkage and density.

Compression tests of geopolymer samples were carried out at the Saga Construction Technology Support Organization (SCTSO) using the concrete testing apparatus in the same test method for concrete samples (sample diameter $d = 50$ mm and height $h = 100$ mm, loading rate $= 0.6 \pm 0.4$ N/mm^2, and loading plate $\varphi = 300$ mm).

8.3.2 Test Results

Compression test results are shown in Fig. 8.2 for key geopolymer samples and Fig. 8.3a, b for mixture geopolymer samples. Figures 8.2 and 8.3a, b show the relationship between the compressive strength q_u and η for $w = w_{\text{opt}} = 0.4$.

8 Chemical and Mechanical Properties of Geopolymers Made …

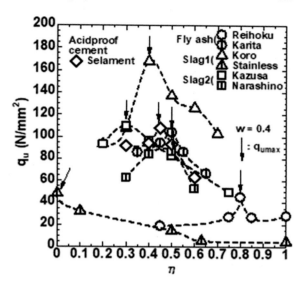

Fig. 8.2 Compression test results for starting geopolymer samples: fly ash, slag1, Slag2 and acidproof cement

As shown in Fig. 8.2 and 8.3a, b, q_u generally increases with a decrease in η and reaches the maximum value q_{umax} at a certain η value, which is defined as the optimum value η_{opt} (Figs. 8.2 and 8.3a, b). The values of η_{opt}, q_{umax} from Figs. 8.2 and 8.3a, b are summarized in Table 8.2. Table 8.2 also includes the value of w_{opt} and both values of the density ρ_t and the volume shrinkage ratio $\Delta V/V$ of geopolymer samples at the q_{umax}.

8.4 Discussion

8.4.1 Correlation Between η_{opt} and Chemical Compositions of Binders

Koumoto (2019) proposed to correlate η_{opt}, which is the optimum value of η yielding q_{umax} and a factor of C_{as} (= $Al_2O_3 + SiO_2$), considering that since the geopolymerization is a chemical reaction, specifically an aluminosilicate reaction, the amount of NaOH which contributes to ionize metals contained in the binder will increase with an increase of the amount of Al_2O_3 and SiO_2.

Figure 8.4 shows the correlation between η_{opt} from Table 8.2 and C_{as} calculated from Table 8.1. In Fig. 8.4 the following equation proposed by Koumoto (2019) is also drawn:

$$\eta_{opt} = 0.0157 C_{as} - 0.414 \qquad (8.1)$$

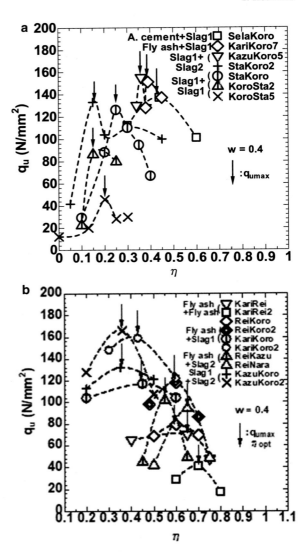

Fig. 8.3 a Compression test results for mixed binder geopolymer samples (Fly ash + Fly ash, Fly ash + Slag1, Fly ash + Slag2 and Slag1 + Slag2). **b** Compression test results for mixed binder geopolymer samples (Acidproof cement + Slag1, Fly ash + Slag1 and Slag1 + Slag2)

The correlation coefficient $r = 0.939$.

Now from Fig. 8.5 which shows the relationship between η_{opt} and each chemical composition of SiO_2, Al_2O_3 and CaO, η_{opt} is shown in proportion to both SiO_2 and Al_2O_3, however, η_{opt} is in inverse proportion to CaO, Although the proportionality between η_{opt} and C_{as} ($= Al_2O_3 + SiO_2$) is considered as proper, η_{opt} should be correlated with C_{as} by including CaO. Here a new factor C_{cas} ($= CaO/C_{as}$) is proposed and the correlation between η_{opt} and C_{cas} is shown in Fig. 8.6. The relationship between η_{opt} and C_{cas} is well correlated and expressed by the following equation with a higher correlation coefficient of $r = 0.950$ than in the case of C_{as} as:

8 Chemical and Mechanical Properties of Geopolymers Made …

Table 8.2 Compression test results for geopolymer samples

Geopolymer materials	Geopolymer samples	q_{umax} (N/mm²)	w_{opt}	H_{opt}	p_t (kg/m³)	AV/V (%)
Fly ash	Reihoku	45.2	0.40	0.80	1,868	0.20
	Karita	104.3	0.40	0.50	1,985	0.20
Slag1	Koro	168.0	0.40	0.40	2,227	1.20
	Stainless	48.1	0.40	0.00	2,269	11.34
Slag2	Kazusa	110.0	0.40	0.30	2,202	1.49
	Narashino	86.4	0.40	0.50	2,043	0.80
Acidproof cement	Selament	108.0	0.40	0.45	2,160	1.10
A. cement + Slag1	SelaKoro	137.0	0.40	0.44	2,218	1.49
Fly ash + Fly ash	KariRei	70.1	0.40	0.65	1,971	0.80
	KariRei2	40.9	0.40	0.70	1,986	1.00
	ReiKoro	79.3	0.40	0.60	2,098	0.70
	ReiKoro2	118.0	0.40	0.60	2,144	1.20
Fly ash + Slag1	KariKoro	117.0	0.40	0.45	2,122	0.90
	KariKoro2	159.0	0.40	0.43	2,149	0.70
	KariKoro7	151.0	0.40	0.39	2,218	1.49
Fly ash + Slag2	ReiKazu	102.0	0.40	0.55	2,081	1.39
	ReiNara	94.3	0.40	0.65	2,024	1.89
	KazuKoro	132.0	0.40	0.35	2,219	1.29
Slag1 + Slag2	KazuKoro2	166.0	0.40	0.36	2,218	1.49
	KazuKoro5	155.0	0.40	0.36	2,249	1.49
	StaKoro	126.0	0.40	0.25	2,243	3.26
Slag1 + Slag1	StaKoro2	136.0	0.40	0.10	2,261	2.68
	KoroSta2	86.6	0.40	0.15	2,296	6.17
	KoroSta5	45.8	0.40	0.20	2,349	5.69

$$\eta_{opt} = 0.842 - 0.580 C_{cas}^{0.855} \tag{8.2}$$

Equation (8.2) is proposed to calculate the η_{opt} value from the value of C_{cas} of binders to effectively produce high compressive strength geopolymers.

8.4.2 Correlation Between q_{umax} and C_{cas}

Figure 8.7 shows the relationship between q_{umax} and C_{cas}. It is of interest that the values of q_{umax} generally show the higher value of 35 N/mm² which is used for the

Fig. 8.4 Correlation between η_{opt} and C_{as}

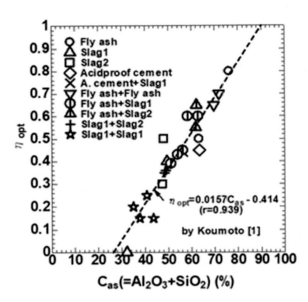

Fig. 8.5 Relationship between η_{opt} and each SiO_2, Al_2O_3 and CaO

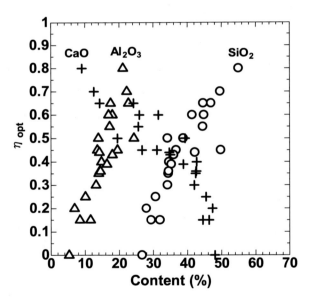

design of the secondary concrete product and that binders having C_{cas} in the range of 0.6–1.0 seems to yield very high compressive strength geopolymers.

Fig. 8.6 Correlation between η_{opt} and C_{cas}

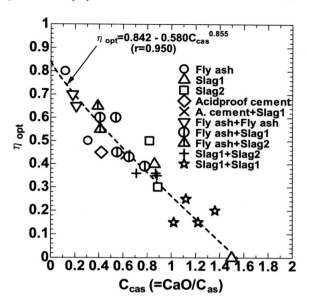

Fig. 8.7 Relationship between q_{umax} and C_{cas}

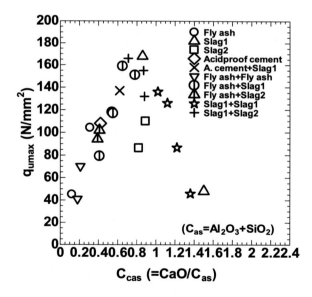

8.4.3 Shrinkage of Geopolymer Samples

Figure 8.8 shows the relationship between the volume shrinkage ratio of the geopolymer sample at q_{umax} $\Delta V/V$ and C_{cas}. The values of $\Delta V/V$ were generally

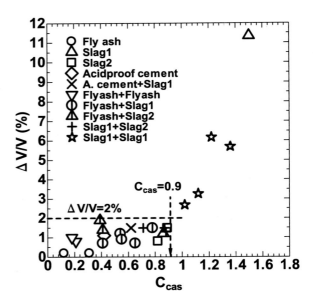

Fig. 8.8 Relationship between $\Delta V/V$ and C_{cas}

less than 2% when C_{cas} is less than 0.9 and increased sharply with an increase in the factor C_{cas}. These results will be helpful for the design of geopolymer construction.

8.4.4 Relationship Between q_{umax} and Density ρ_t for Geopolymer Sample and Natural Rock

Figure 8.9 shows the relationship between q_{umax} and densities ρ_t of geopolymer samples and natural rocks (Kojima and Nakao 1995). From Fig. 8.9 although the values of ρ_t of geopolymers are smaller than that of natural rocks, the values of q_{umax} of geopolymers are not so much smaller than that of natural rocks. It is noted that although natural rocks have been made under high temperature or very high pressure or both for a long curing period, geopolymers are made simply by a chemical reaction called geopolymerization under room temperature and a short curing period (28 days).

8.5 Conclusions

This paper describes the chemical and mechanical characteristics of geopolymers by mixing fly ash and slags as binders with 48% NaOH (18 mol/L) and sodium silicate ($Na_2 \cdot nSiO_2$, $n = 3.2$) as activators. The research results are summarized as:

8 Chemical and Mechanical Properties of Geopolymers Made ...

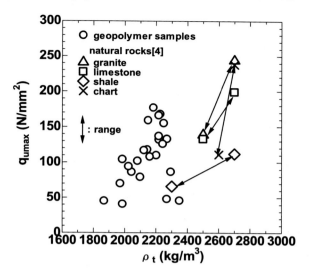

Fig. 8.9 Relationship between q_{umax} and ρ_t of geopolymer samples and natural rocks

(a) There is an optimum value, η_{opt} (the weight ratio of NaOH to sodium silicate) yielding q_{umax} for each binder,

(b) The η_{opt} is well correlated with the factor C_{cas} (=CaO/(Al$_2$O$_3$ + SiO$_2$)) as: $\eta_{opt} = 0.842 - 0.580 C_{cas0}^{855}$ (see Fig. 8.6) and the value η_{opt} for an arbitrary binder necessary to manufacture the high compressive strength geopolymer is calculated by the above equation,

(c) The values of q_{umax} generally show the higher than the value of 35 N/mm^2 which is used for the design of the secondary concrete product for C_{cas} in the range of 0.6–1.0,

(d) The volume shrinkage ratio $\Delta V/V$ is generally less than 2% when C_{cas} is smaller than 0.9,

(e) The compressive strength of geopolymers is generally similar to that of natural rocks although the densities of geopolymers are lower than that of natural rocks.

Acknowledgements The author is grateful to Maruwa Giken for preparing ground solid waste incinerator slags. The author is also grateful to Kyuden Sangyo, Shunan Works, Nissin Steel, and Nippon Steel & Sumikin Slag Products for contributing coal-fired power plant ash, stainless manufactured slag, and ground granulated blast slag, respectively.

References

Buchwald A (2006) What are geopolymers? Current state of research and technology, the opportunities they offer, and their significance for precast industry. Betonwerk Fertigteil Tech 72(7):42–49

Davidovitz J (1991) Geopolymers, inorganic polymeric new materials. J Therm Anal 37(8):1633–1656. https://doi.org/10.1007/BF0.1912193

Kojima K, Nakao K (1995) Ground technology basics and practice. KAJIMA INSTITUTE PUBLISHING CO., LTD, p 39

Koumoto T (2019) Production of high compression strength geopolymers considering fly ash or slag chemical composition. J Mater Civ Eng ASCE 31(8)

Chapter 9
Characteristics of Re-liquefaction Behaviors of the Typical Soils in the Aso Area of Kumamoto, Japan

Guojun Liu, Noriyuki Yasufuku, Ryohei Ishikura, and Qiang Liu

9.1 Introduction

An earthquake with two great shocks, the first shock at $M_w = 6.2$, and the second shock at $M_w = 7.0$, struck the Kumamoto area, Japan in 2016. The earthquakes induced several soil liquefaction related disasters in vast areas. Such that many landslides, numerous foundations of residential buildings and infrastructures were destroyed during the earthquakes, which was traceable to soil liquefaction. These disasters threatened local human lives and economic activities severely (Mukunoki et al. 2016). Moreover, several investigations indicated that the disaster extents were altered greatly, after the foreshock-mainshock series. Murakami et al. (2018) and Hirata et al. (2018) compared the differences in liquefaction sites between the two shocks, by aerial photographs from several areas in this region; Wakamatsu et al. (Wakamatsu et al. 2016) arranged sites where the sand boiling appeared between the two shocks as illustrated in Fig. 9.1. The researchers indicated that both the affected areas and severity expanded significantly after the principal shock (second shock).

Besides, the geological condition of the Kumamoto-Aso area was influenced by Aso Volcano significantly. Owing to the eruption of Aso in history, volcanic ash was deposited in this area, to produce a unique volcanic soil with a mass of black fines. Photo 9.1 presented a sand boiling, during the earthquakes. Regarding these,

---

G. Liu (✉)
Changshu Institute of Technology, No.99 South Third Ring Road, Changshu, Jiangsu 215500, China
e-mail: liuguojun101@gmail.com

N. Yasufuku · R. Ishikura
Kyushu University, 744 Motooka Nishi-ku, Fukuoka 819-0395, Japan

Q. Liu
Shandong University of Science and Technology, 579 Qianwangang Road, Huangdao District, Qingdao, Shandong 266590, China

© The Author(s), under exclusive license to Springer Nature Singapore Pte Ltd. 2023
H. Hazarika et al. (eds.), *Sustainable Geo-Technologies for Climate Change Adaptation*, Springer Transactions in Civil and Environmental Engineering,
https://doi.org/10.1007/978-981-19-4074-3_9

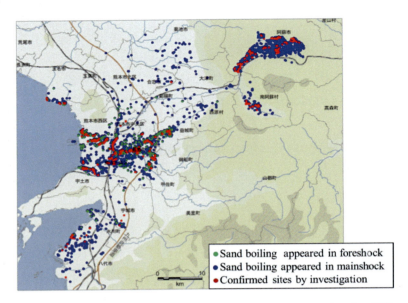

Fig. 9.1 Liquefaction sites by in-situ investigation of 2016 Kumamoto Earthquakes (Wakamatsu et al. 2016)

Photo 9.1 Appearance of sand boiling in 2016 Kumamoto Earthquakes (Nakano et al. 2017)

it is crucial to understand that the re-liquefaction characteristics in the earthquakes depended on the unique soil conditions in the Kumamoto-Aso area.

9.2 Test Conditions

9.2.1 Test Materials

To investigate the liquefaction and re-liquefaction behaviors, two typical soils were collected, considering the local geological conditions. The first soil was collected from the site where sand boiling appeared in the earthquakes, as illustrated in Photo 9.1. This soil contains several black fines contributed by volcanic ash. These black fines were greatly considered to potentially affect the liquefaction behaviors in the earthquakes (Murakami et al. 2017). To clarify this, a pure volcanic soil named Kuroboku was collected near the Aso Volcano, as the test material as well. In addition, Toyoura sand was selected as a typical fine sand, and it played the role of the contrast materials in this study.

The grain size distributions of these test materials were arranged in Fig. 9.2. Kuroboku was collected from a landslide in the earthquakes near the Aso Volcano. To obtain enough test material, Kuroboku was taken from sites A and B which were close to each other. The figure indicated that all the soils were distributed in the silt-sand zone majorly. Kuroboku contained the highest fines, meanwhile, the particles were less than 0.2 mm mostly. The soil that erupted in the sand boiling site presented a wider distribution than the other materials. It contained both the finer granules, and the coarser granules than Toyoura sand. However, Kuronoku contained times of fine (≤0.075 mm), in comparison with the soil that erupted in the liquefaction site. To restrain the influence from the fines, the fines in Kuroboku were reduced to compare the behaviors with the erupted soil in the study. Toyoura sand as a material prone to

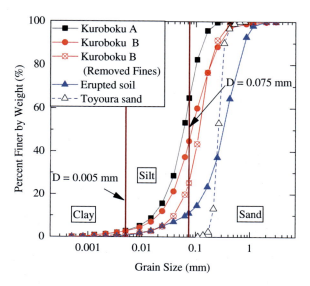

Fig. 9.2 Cumulative grain size distribution of test materials

Table 9.1 Basic physical properties for the typical soils related to liquefaction disasters in the 2016 Kumamoto Earthquakes

Sample	Specific gravity, G_s	Max. void ratio, e_{max}	Min. void ratio, e_{min}	Liquid limit, LL (%)	Plastic limit, PL (%)	Plastic index PI
Kuroboku A	2.390	None	None	212.5	146.3	66.2
Kuroboku B	2.320	None	None	157.3	112.1	45.2
Kuroboku B removed fines	2.320	2.351	1.594	59.3	57.2	2.0
Boiled sand	2.755	1.618	0.951	NP	NP	NP
Toyoura	2.643	0.977	0.606	NP	NP	NP

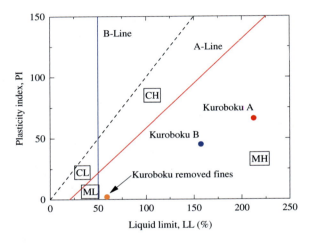

Fig. 9.3 Plasticity Chart of Kuroboku A and B, and Kuroboku B removed fines

liquefy, was taken into the tests as a judgment for the liquefaction behaviors of the other materials. More properties of the test materials were presented in Table 9.1 and Fig. 9.3.

9.2.2 Test Apparatus and Program

Cyclic tri-axial compaction apparatus was adopted for the tests in the study. The pure Kuroboku was set to the specimens at similar void ratios in each case. The specimens of Toyoura sand, and erupted soil from the sand boiling site were prepared with the relative density, at $D_r = 60$–80%. The specimens were well saturated in each test. Further, the specimens were applied by an adequate consolidation process, under drainage conditions. Subsequently, the cyclic load tests were started with a frequency of 0.1 Hz, until the axial strain with double amplitudes DA is 5%. Until now, the progress was called the first liquefaction process in this study. Further, the specimens were repeated after the consolidation process, and after the cyclic load test as well. This progress was called the second liquefaction process. The confining pressure was applied isotropically in each consolidation process.

9.3 Tests Results and Discussions

9.3.1 Liquefaction Behaviors of Pure Volcanic Soil

Three cases were tested for the Kuroboku, which was unabridged for the grain size distribution. The content of fines was measured to be approximately 65% by weight. In addition, it was classified as the silt with a very high liquid limit. The behaviors were presented in Fig. 9.4. It is clearly visible that Kuroboku is a very difficult material to liquefy.

In C-3, the specimen was applied to the cyclic loading, until the axial strain with amplitude in double directions reached DA = 5%, while the excess pore pressure $u/\sigma c_0' = 0.842$. Further, the sample was consolidated again with drainage. The void ratio increased in level from 1.28 to 1.14, which also led to a negligible increase in the cycle stress ratio. Subsequently, the cyclic loading was applied again to trigger DA = 5% once, and the test was terminated. It could be found that the necessary cycles increased significantly from 625.5 to 823, even if the cyclic stress ratio increased. Similar characteristics could be discovered in C-1. The cyclic resistance (at DA = 1.35) increased as well when the sample was pre-sheared with DA = 1.54%. The results indicated that the cyclic resistance would increase by the pre-shearing within DA = 5%, because of the existence of fines.

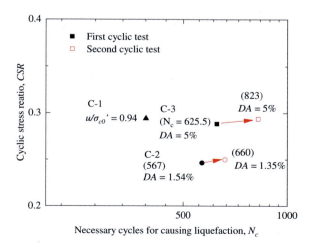

Fig. 9.4 Cyclic resistance of Kuroboku by different loading conditions

When the sample is detected to determine the possible maximum excess pore pressure, such as C-2 in this study, a significantly large deformation occurs. The axial strain with amplitude in double directions increased to 25.68%. A necking appeared in the middle of the specimen in height. It cannot recover when unloaded, leading to unavailability for the second test stage. Therefore, natural pure Kuroboku exhibited an un-liquefiable feature owing to its great fine content.

9.3.2 Liquefaction Behaviors of Kuroboku Removed Fines

The fines were reduced in Kuroboku to obtain a similar fine content similar to the erupted soil from the sand boiling site in the earthquakes. However, it was very difficult to separate the fine particles, approximately 25%, still contained in the tests. The specimens were prepared with relative density of approximately 60%, 70%, and 85% respectively.

The samples were divided into two groups. The first one, including four cases, which were prepared with a relative density of approximately 60%, was supposed to investigate the general behavior of liquefaction and re-liquefaction. The other two cases were adopted to find the effects of different relative densities. The cyclic test results were arranged by DA = 5%, as illustrated in Fig. 9.5.

Approximately 10% in relative density increased after the first liquefaction stage and re-consolidation, notwithstanding the samples being prepared with an initial relative density of approximately 60, 70, or 85%. Simultaneously, the liquefaction resistance also increased significantly with increasing relative density. Regardless of the first or second stages, the liquefaction resistance was significantly affected by the relative density. It was also discovered that the failure states of the sample with the initial relative density of 70 and 85%, were very close to the failure line for the

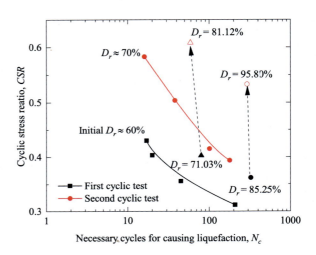

Fig. 9.5 Liquefaction and re-liquefaction behaviors of Kuroboku removed fines

sample with initial relative density $D_r = 60\%$ in the second cyclic test stage, while it increased to approximately 70%. Whether these results appeared occasionally in this study or not, requires more test results and discussion in the future.

9.3.3 Liquefaction Behaviors of the Erupted Soil from the Sand Boiling Site

The specimens of the erupted soil were prepared to two relative densities at approximately 60 and 80%. As these results were illustrated in Fig. 9.6, the liquefaction resistance for the first test stages was not affected basically, by the relative density between 60 and 80%. The failure lines were very close. This behavior could also be extended to the performance in the second test stage. The liquefaction resistance in the second stages was raised, however, the difference was negligible. The resistance could be viewed without changes in general consideration. In addition, the liquefaction resistance in the second stages decreased, with decreasing cyclic stress ratio. It would become less than in the first stages when the cyclic stress ratio decreased to lower than 2.0 for the sample with $D_r = 60\%$.

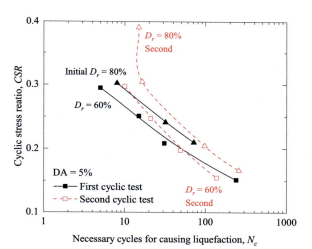

Fig. 9.6 Liquefaction behaviors of boiled sand in liquefaction and re-liquefaction tests

9.3.4 Comparison of Liquefaction Behaviors Between Toyoura Sand, Kuroboku, and the Erupted Soil from the Sand Boiling Site

To investigate the liquefaction behavior of the local soils in the Kumamoto-Aso area, especially for detecting the potential influence of the volcanic particles which were deposited in this area, it is worth comparing the liquefaction behavior between Toyoura sand, Kuroboku, and boiled sand. Toyoura sand was regarded as the base for discovering the relative differences in liquefaction potential. The failure line of DA = 5% was adopted here. The result of Toyoura sand was compared with Kuroboku removed fines in Fig. 9.7, and compared with boiled sand in Fig. 9.8.

On the other hand, although the boiled sand as a soil was considered to contain the volcanic particles, the test results indicated a very low liquefaction resistance compared to Toyoura sand, even if there was approximately 10% of fines contained. It may be because the exact void is significantly greater than Toyoura sand. The void ratio was tested from 1.618 to 0.951 for boiled sand, and from 0.977 to 0.606 for Toyoura sand. The boiled sand was always retained, with a void ratio of approximately 2 times as Toyoura sand. The change in the liquefaction resistance was also negligible. This slight increase depended on the complete drainage, and enough consolidation after the first liquefaction. In real disasters, such as the 2016 Kumamoto earthquakes, the two shocks appear within a short interval, and the liquefaction resistance might be reduced furtherly in the ground.

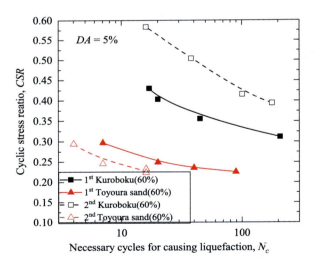

Fig. 9.7 Comparison of re-liquefaction behaviors between Toyoura sand and Kuroboku

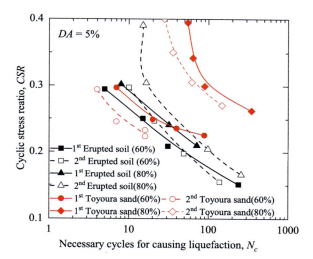

Fig. 9.8 Comparison of liquefaction behaviors between Toyoura sand and erupted soil

9.4 Conclusions

Based on the recurrence of liquefaction in the 2016 Kumamoto Earthquake, the two typical local soils were tested, and their basic physical properties and liquefaction behaviors were investigated. The erupted soil from the sand boiling site in the earthquakes, was selected to represent the typical ground condition of the Kumamoto-Aso area. Kuroboku, a representative volcanic soil in Aso-Kumamoto Area, was selected to discuss the effects of the volcanic grains contained in the local ground condition. Through the tri-axial cyclic tests, the liquefaction and re-liquefaction behaviors of these typical soils were detected by comparing them with fine sand in this study. The results were concluded as follows:

1. Natural pure volcanic soil, Kuroboku, is considered as an un-liquefiable soil.
2. The erupted soil is prone to liquefaction because of its higher void ratio in comparison with fine sand.
3. The fine volcanic ash affects the liquefaction behaviors of both Kuroboku and the erupted soil significantly. It provides a positive effect on the liquefaction resistance of local soils.
4. The liquefaction resistance of the erupted soil was strengthened after the first liquefaction process and the re-consolidation process. However, the resistance in the second liquefaction process is insignificant, because of its low absolute liquefaction resistance.
5. Owing to the local ground conditions, the soil liquefaction will appear repeatedly in the earthquakes with a series of shocks. The disaster will be worse if the series of shocks strike here during the short intervals.

References

Hirata R, Murakami S, Hashihara H, Nomiyama Y (2018) Investigation of liquefiable layer in liquefaction zone in the H28 Kumamoto earthquakes. In: Proceedings of the 73rd JSCE annual conference, Sapporo, Japan, Japanese, pp 83–84

Mukunoki T, Kasama S, Murakami H, Ikemi H, Ishikura R, Fujikawa T, Yasufuku N, Kitazono Y (2016) Reconnaissance report on geotechnical damage caused by an earthquake with JMA seismic intensity 7 twice in 28 h, Kumamoto, Japan. Soils and Found JGS 56(6):947–964

Murakami S, Nagase H, Osato S, Yakabe H (2017) Chapter 5: investigation of liquefaction and ground settlement. In: JGS (eds) Report on the investigation of geological disasters in the H28 Kumamoto earthquakes, JGS, Japanese, pp 115–141

Murakami S, Hirata R, Tanabe H, Miwa S (2018) Effects of foreshock and mainshock on the expansion of three liquefaction zones in the H28 Kumamoto earthquakes. In: Proceedings of the 73rd JSCE annual conference. JSCE, Sapporo, Japan, Japanese, pp 81–82

Nakano T, Liu H, Nagase H, Hirooka A, Tomohisa T (2017) Study on the characteristic of the soils from sand boil in the Kumamoto earthquakes. In: Proceedings of JSCE Western branch annual meeting, Fukuoka, Japan, Japanese, pp 371–372

Wakamatsu K, Senna S, Ozawa K (2017) Liquefaction and its characteristics during the 2016 Kumamoto Earthquake. J Jpn Assoc Earthq Eng 17(4):81–100 (Japanese)

Part III
Sustainable Development for Infrastructures

Chapter 10
Sustainable Transport Infrastructure Adopting Energy-Absorbing Waste Materials

Buddhima Indraratna, Yujie Qi, and Trung Ngo

10.1 Introduction

By 2018, the Australian rail network had more than 36,000 km of track and ranked seventh highest worldwide (Indraratna et al. 2012a). Until recently most of these tracks had been built with ballast, but ballasted tracks deteriorate progressively as the ballast breaks due to particle splitting and attrition and abrasion from dynamic loading during service. This process of degradation will exacerbate because faster and heavier trains are needed to cope with the growing population and increasing freight movements. In addition, the impact loading caused by rail and/or wheel imperfections and irregularities also inevitably intensifies ballast degradation (Indraratna et al. 2019), and this will lead to costly and more frequent track maintenance. It is, therefore, imperative to improve the design of rail track and concentrate more on the track substructure, including ballast, subballast and subgrade layers, to solve the geotechnical problem of ballast degradation.

B. Indraratna (✉)
Director of Transport Research Centre (TRC) and Founding Director of Australian Research Council's Industrial Transformation Training Centre for Advanced Technologies in Rail Track Infrastructure (ITTC-Rail), Faculty of Engineering and Information Technology, University of Technology Sydney, Sydney, NSW 2007, Australia
e-mail: Buddhima.indraratna@uts.edu.au

Y. Qi · T. Ngo
Transport Research Centre, Faculty of Engineering and Information Technology, University of Technology Sydney, Sydney, NSW 2007, Australia

© The Author(s), under exclusive license to Springer Nature Singapore Pte Ltd. 2023
H. Hazarika et al. (eds.), *Sustainable Geo-Technologies for Climate Change Adaptation*, Springer Transactions in Civil and Environmental Engineering,
https://doi.org/10.1007/978-981-19-4074-3_10

The adoption of artificial geo-inclusions such as resilient rubber materials, geogrids and geocomposites has already proved to be a very efficient method of mitigating the degradation and deformation of ballast (Indraratna et al. 2005, 2012b, 2014a, b, 2016, 2017a, b, c, 2018, 2019, 2020; Ngo et al. 2018, 2019; Jayasuriya et al. 2019; Nimbalkar and Indraratna 2016; Navaratnarajah et al. 2018; Qi et al. 2018a, b, c, d, 2019a, b; Sol-Sánchez et al. 2015; Esmaeili et al. 2017; Signes et al. 2015; Navaratnarajah and Indraratna 2017; Qi and Indraratna 2020). In recent years, the use of recycled rubber products such as rubber mat/pats, tyre cells, and granulated rubber has prevailed over other materials due to their high energy absorbing capacity and high damping property (Qi et al. 2018a, b, c, 2019a, b; Indraratna et al. 2017b; Sol-Sánchez et al. 2015). Esmaeili et al. (2017) found that adding 5% of tyre derived aggregates could reduce the breakage of fouled ballast by 34%, while Signes et al. (2015) proved that mixing 1–10% of rubber particles with subballast helps rail tracks resist degradation, and Indraratna et al. (2014a) found that installing under ballast mats at the ballast-deck interface attenuates the impact of dynamic train load imparted by the running stock.

Since there is no comprehensive scientific evidence on ballast degradation and its associated load-deformation responses under cyclic and impact loading when recycled rubber materials are incorporated in track embankment, this paper reviews current research in this area carried out by the researchers of the current Transport Research Centre (TRC) currently at the University of Technology Sydney (UTS) over the past decades. This research consists of a series of prototype cubical triaxial tests and drop hammer impact tests to examine the performance of under sleeper pads (USP), under ballast mats (UBM), rubber energy absorbing drainage sheets (READS), tyre cell reinforced track and a synthetic energy absorbing layer (SEAL; a matrix of rubber crumbs and mining rejects). It also consists of finite element modelling (FEM) for track reinforced with tyre cells and field tests for shock mats (UBM) with geogrids.

10.2 Large-Scale Laboratory Tests

10.2.1 Large-Scale Cubical Triaxial Tests

Testing Facility and Sample Preparation. Large-scale cubic triaxial tests were conducted using the prototype track process simulation testing apparatus (TPSTA) as shown in Fig. 10.1a. It has a machine chamber that is 600 mm deep by 800 mm long by 600 mm wide; these dimensions duplicate a unit cell of the Australian standard track. The four sidewalls are allowable to move to simulate the lateral deformation in the longitudinal and transverse direction of the track. In this study, longitudinal deformation of the track is assumed to be negligible (i.e. plain strain condition), so during the test the sidewalls parallel to the sleeper were locked in position by the hydraulic systems.

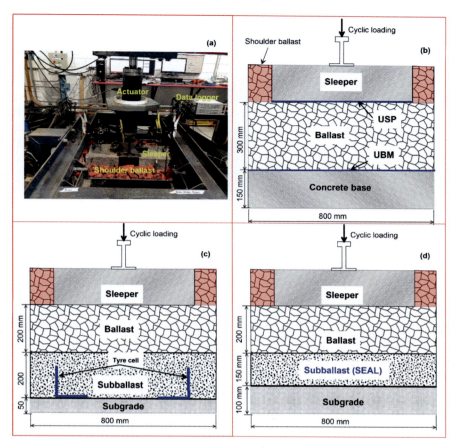

Fig. 10.1 **a** Cubical cyclic triaxial apparatus; schematic illustrations of **b** test specimens with USP or UBM, **c** tyre-cell reinforced track specimen and **d** test specimen incorporated SEAL. Modified after Indraratna et al. (2020)

The test specimens with either rubber mats (UBM) or rubber pads (USP) were tested with stiff subgrade condition to simulate tracks built in a tunnel or on a bridge deck. Therefore, the test specimen was prepared with a ballast layer (depth: 300 mm) sitting on a 150 mm thick concrete base (Fig. 10.1b). The USP (790 × 200 × 10mm) and UBM (790 × 590 × 10mm) were made from recycled tyres by removing the steel cords and fibre, and Fig. 10.1b shows their positions in the test specimen. The test sample reinforced with a tyre cell was prepared with three layers, a 200 mm thick layer of ballast, a 200 mm thick layer of subballast filling in a tyre cell, and a 50 mm thick layer of subgrade (Fig. 10.1c). The tyre cell was made from the recycled tyres by removing one sidewall. The test specimen incorporating SEAL has also been compacted in three layers as shown in Fig. 10.1d. The novel subballast i.e. SEAL was a mixture of recycled rubber crumbs (RC) and mining rejects, i.e. coal wash (CW) and steel furnace slag (SFS). The SEAL matrix was prepared with different amounts of RC and SFS: CW = 7:3 mixed by mass suggested by Indraratna et al. (2017a) and Qi et al. (2018d) to avoid unacceptable particle degradation of CW and volumetric expansion of SFS while maintaining sufficient strength of the waste SEAL matrix.

All cyclic triaxial tests were running with a loading frequency of $f = 15$ Hz to simulate the train speed of 115 km/h (Indraratna et al. 2017b). The specimen reinforced with a tyre cell was tested under the maximum vertical stress of 385 kPa to simulate a heavy haul freight train having a 40-tonne axle load, while the remaining tests had a lower vertical pressure of 230 kPa for a normal 25-tonne axle load train. All the cyclic loading tests were completed when 500,000 cycles were achieved. Details of how the specimen was prepared and tested can be found elsewhere (Jayasuriya et al. 2019; Indraratna et al. 2017b; Navaratnarajah and Indraratna 2017; Qi and Indraratna 2020).

Vertical and Lateral Displacement. Test results for vertical and lateral deformation changing with loading cycles of the test specimen are shown in Fig. 10.2. All the test specimens settled quickly within the first several thousand cycles caused by particle densification and ballast breakage, and then gradually stabilised after 100,000 cycles. Note that with stiff subgrade the addition of rubber mats/pads (i.e. UBM, USP) reduces the settlement and lateral displacement of ballast, but the addition of USP is more efficient than UBM (Fig. 10.2a, b). Having a tyre cell in the subballast layer reduces settlement by 10–12 mm under a heavy track loading condition compared to the unconfined test specimen (Fig. 10.2c). The lateral deformation of the test specimen reinforced with a tyre element shows contraction, whereas the test specimen without a tyre cell is dilative (Fig. 10.2d). This is mainly because tyre cells included in the subballast layer increase the confining pressure on the specimen and prevent the particles from moving outward (Indraratna et al. 2017b). Figure 10.2e, f shows the settlement and lateral movement of a specimen with SEAL. It can be seen that increasing the amount of RC inside SEAL the settlement increases, but lateral dilation decreases when RC < 20% and there was a large lateral fluctuation when the RC ≥ 20%. The specimen with 40% RC in SEAL failed within 1500 cycles due to excessive vibration and settlement (Qi and Indraratna 2020). This test result indicates that a certain amount of RC (10%) in SEAL will help to reduce the lateral dilation and vertical displacement of a traditional track specimen, whereas too much RC included may induce track instability (e.g. extensive vibration and settlement).

Ballast Degradation. After each test, the ballast compacted directly under the sleeper was separated collected to check the particle size distribution (PSD) curve to evaluate ballast degradation using the ballast breakage index (BBI) which is initially developed by Indraratna et al. (2005). The BBI is defined in Fig. 10.3a–d shows BBI for the specimen with different rubber inclusions under different test conditions. It is obvious that BBI decreases as the damping rubber products (i.e. UBM, USP, tyre cell or rubber crumbs) are incorporated in the track specimen regardless of the test conditions. As with the response of vertical deformation, the inclusion of USP is more effective than UMB, as USP reduces the BBI of the track specimen by 60%, while the UBM reduces the BBI by 24%. Ballast breakage under heavy haul loading conditions is more severe with BBI = 0.2, but when confined by a tyre cell the BBI

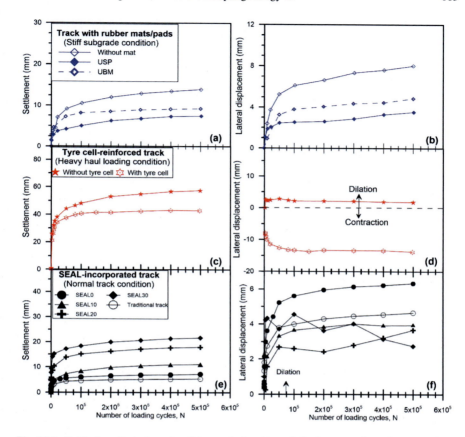

Fig. 10.2 Cubical cyclic test results of the vertical and lateral displacement for **a, b** test specimen with and without rubber mats/pads, **c, d** tyre cell reinforced track specimen and **e, f** test specimen incorporated SEAL. Data sourced from Jayasuriya et al. (2019), Indraratna et al. (2017b), Navaratnarajah and Indraratna (2017), Qi and Indraratna (2020)

decreases significantly to 0.063. The BBI of the track specimen that incorporates SEAL without RC is similar to the traditional track specimen, but it decreases by more than half when 10% of RC is included in the SEAL matrix. However, when more RC (>10%) is added to SEAL, the BBI does not reduce more, which suggests that 10% RC in SEAL is enough to mitigate ballast degradation.

10.2.2 Drop Hammer Impact Tests

Test Facility and Sample Preparation. The dynamic impact force at the transition zones (between soft and stiff subgrade conditions) of the ballasted track is one of the

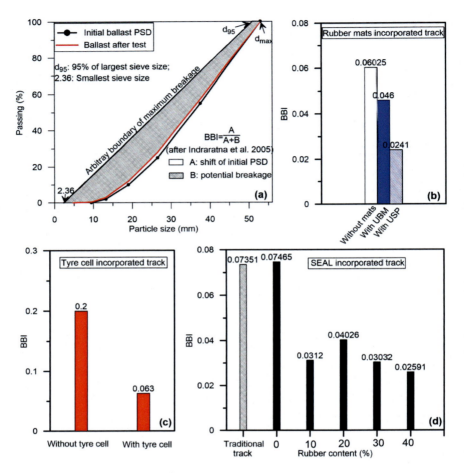

Fig. 10.3 a Definition of the ballast breakage index (BBI); BBI of **b** track specimens with USP/UBM, **c** tyre cell confined track specimens and **d** SEAL incorporated track specimen. Data sourced from Jayasuriya et al. (2019), Indraratna et al. (2017b), Navaratnarajah and Indraratna (2017), Qi and Indraratna (2020)

key factors that induce the track instability and ballast degradation. Ngo et al. (2019) proposed a solution that using a rubber energy absorbing drainage sheet (READS) placing underneath the ballast layer to attenuate the impact force and reduce ballast degradation. They then examined its efficiency through a series of large-scale impact loading tests using the drop hammer impact apparatus shown in Fig. 10.4a. The test sample was prepared within a rubber membrane, and the schematic illustration of the test specimen is shown in Fig. 10.4b where a 10 mm-thick READS made from recycled rubber was installed between the ballast layer (depth: 350 mm) and the sub-ballast layer (depth: 100 mm). The 50 mm thick layer of subgrade was either soft or stiff (concrete base). Each test took 15 drops from 100–250 mm high to simulate an

10 Sustainable Transport Infrastructure Adopting Energy …

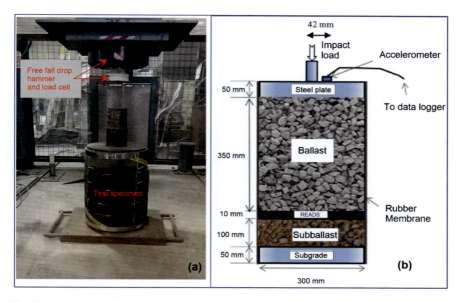

Fig. 10.4 **a** Drop weight impact apparatus with a prepared test specimen; **b** schematic illustration of the test specimen for the impact test. Modified after Ngo et al. (2019)

impact force between 250 and 550 kPa. Details of the test procedures and materials can be found elsewhere (Ngo et al. 2019).

Test Results. The final vertical and lateral ballast deformation of test specimens subjected to impact force with and without READS is shown in Fig. 10.5a–d. Note that the vertical and lateral movement increase with the drop height (h_d, mm) regardless of the subgrade conditions and they decrease when READS are placed under the ballast. Without READS the vertical and lateral deformation of the test specimen with a stiff subgrade is similar to those with a soft subgrade. To better evaluate the efficiency of using READS under different subgrade conditions, the percentage reduction in vertical and lateral deformation and breakage is shown in Fig. 10.5e, f.

The percentage reduction factor (%) in vertical displacement (R_v), in lateral deformation (R_h), and in ballast breakage index (R_b) can be calculated as:

$$R_v = (1 - \frac{S_{v(\text{WithREADS})}}{S_{v(\text{NoREADS})}}) \times 100) \quad (10.1)$$

$$R_h = (1 - \frac{S_{h(\text{WithREADS})}}{S_{h(\text{NoREADS})}}) \times 100 \quad (10.2)$$

$$R_b = (1 - \frac{S_{b(\text{WithREADS})}}{S_{b(\text{NoREADS})}}) \times 100 \quad (10.3)$$

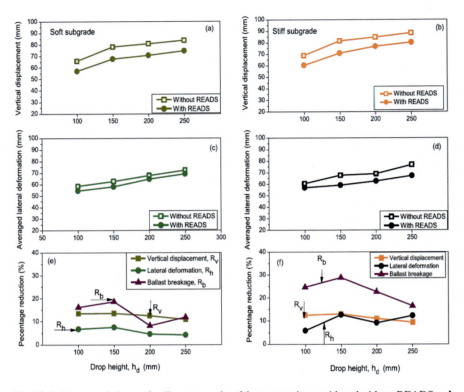

Fig. 10.5 Large-scale impact loading test results of the test specimen with and without READS: **a, b** vertical deformation, **c, d** average lateral deformation and **e, f** percentage reduction for deformation and ballast degradation (Ngo et al. 2019)

where S_v, S_h, and S_b are the vertical displacement, horizontal (lateral) deformation and BBI of the test specimen. It is obvious that the reduction in BBI and lateral deformation by adding READS is more pronounced with a stiff subgrade, whereas the reduction in vertical deformation is comparable for both types of subgrade. Note also that READS can help to reduce the vertical deformation by approximately 7–15% and BBI by up to 28% on the stiff subgrade. These results corroborate the energy absorbing concept for using rubber inclusions that when more energy is absorbed in the resilient rubber layer, the energy exerted onto the ballast and other track substructure layers will be attenuated and thus reduce ballast breakage and track deformation (Qi et al. 2018a; Qi and Indraratna 2020).

10.3 Finite Element Modelling of Using Waste Tyres in Tracks

Scrap tires are a major environmental concern because every year Australia produces more than 50 million waste tires, and since only 13% are recycled the remainder goes into landfill or illegal dumping. Recent research at the University of Wollongong Australia (UOW) indicates that waste tires installed under track foundations (Fig. 10.6a) will increase track bearing capacity and reduce lateral displacement. A three dimensional finite element analysis (3D-FEM) using ABAQUS is implemented (Fig. 10.6b) to examine the induced stress-displacement response of a sub-ballast layer that consists of infilled rubber tyres. For the sake of brevity, further details about the model setup, and the loading and boundary conditions can be found in Indraratna et al. (2017c).

Figure 10.6c shows how rubber tyres reduce the deviator stress on the subgrade. It is predicted that the maximum deviator stress happens at the ends of sleepers and decrease towards the middle of the track. The FEM simulations show that a tyre reinforcement assembly has reduced stress by almost 12% compared to an unreinforced track section (for a 25-tonne axle load, speed: 90 km/h). Essentially, the additional confinement supplied by the tyres stiffens the gravel-infilled composite aggregates and allows more uniform stress to be transmitted to the underlying layers. Figure 10.6d shows the typical contours of horizontal displacement predicted for the

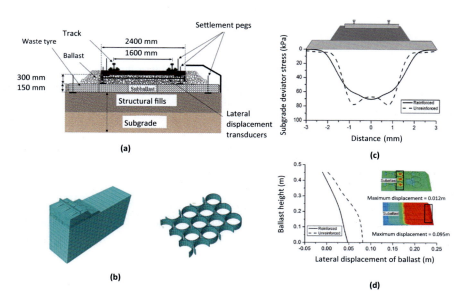

Fig. 10.6 **a** Typical track dimensions with rubber tyres reinforced capping layer; **b** FEM mesh for tracks and the tyre assembly; **c** stress distributions below reinforced and unreinforced track; **d** simulated lateral displacements. Modified after Indraratna et al. (2017c)

sub-ballast; here the highest lateral deformation of the unreinforced track is higher than the reinforced case because the additional lateral confinement provided by the tyre stiffens the infill-aggregate assembly enables a more uniform load to be transferred to the underlying subgrade; as a consequence, there is reduced lateral displacement. The results from large-scale laboratory tests and numerical modelling indicate that placing recycled tyres under the ballast layer will greatly improve overall track stability.

10.4 Field Test

To investigate how different geo-inclusions and shock mats affect track performance, a field case study was carried out on fully instrumented tracks at Singleton, NSW Australia (Fig. 10.7a). This track belongs to the Australian Rail Track Corporation (ARTC).

Construction of Tracks. Eight sections of instrumented track were built on different types of subgrades, (i) relatively soft soils mixed with alluvial-silty clay deposits, (ii) stiff reinforced concrete bridge decks (Mudies Creek), and (iii) siltstone based subgrade. A typical track substructure was built on a 300 mm thick ballast followed by a 150 mm thick sub-ballast (capping). A 500 mm thick structural fill layer was laid above the subgrade. Three types of geogrids and geo-composite (geogrid + non-woven geotextiles) were installed below the ballast layer. The biaxial geogrids include: (1) geogrid_G1 (aperture: 44 × 44 mm, tensile strength: 36 kN/m); (2) geogrid_G2 (aperture: 65 × 65 mm, tensile strength: 31 kN/m); and (3) geogrid_G3 (aperture: 40 × 40 mm, tensile strength: 30 kN/m). A geo-composite layer (aperture size: 31 × 31 mm, tensile strength: 40 kN/m) attached to a non-woven polypropylene geotextile (weight: = 150 g/m^2, thickness = 2.9 mm). A rubber mat (shock absorbing mat) was placed under the ballast at the bridge-deck to eliminate ballast breakage (Fig. 10.7b). More details on the engineering properties of these materials are contained in Indraratna et al. (Indraratna et al. 2014b).

Track Instrumentation and Measurement. Different sensors and instrumentations were used to record data, as shown in Fig. 10.7c. Strain gauges installed on the grid were used to measure mobilized strains. Dynamic stresses induced by train passage were recorded by 200 mm diameter pressure plates, and the vertical deformation of ballast was measured by electronic potentiometers (Fig. 10.7d). Settlement pegs were also used to capture the vertical displacement of the ballasted tracks. Electrical analogue signals from instrumentation and sensors were imparted using a mobile data logging system at a frequency of 2000 Hz (Fig. 10.7e). More details of the instrumentation, the data acquisition process, and the layout of sections have been described earlier by Indraratna et al. (2014b, 2016).

Measured settlements. Figure 10.8 shows the vertical settlement S_v of ballast measured with increased load cycles, N recorded for soft subgrades and concrete bridge (hard rock). These results indicate that the increasing rate of S_v decreases as

10 Sustainable Transport Infrastructure Adopting Energy … 169

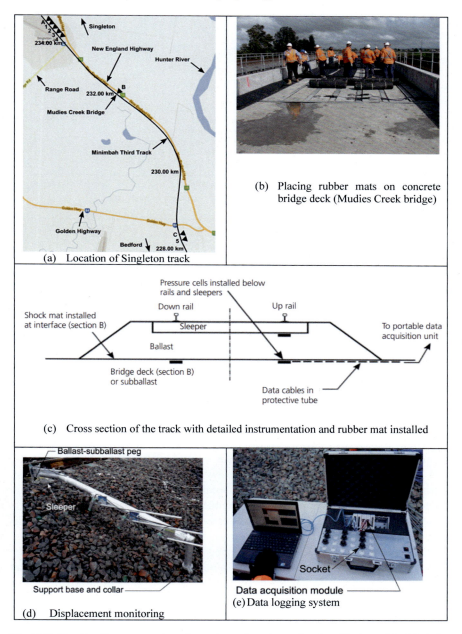

Fig. 10.7 Details of instrumentation of experimental sections of track at Singleton. Modified after Indraratna et al. (2014b)

Fig. 10.8 Measured settlements of ballast for **a** soft subgrade; **b** hard rock. Data source Indraratna et al. (2014b)

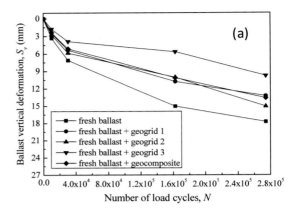

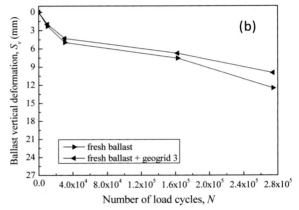

the number of load cycles N increases. The S_v for the reinforced tracks is almost 10–30% less than unreinforced tracks, proving how geogrid has reduced track settlement. A similar pattern is also observed in the laboratory (Indraratna et al. 2012b), mainly due to the particle-geogrid interlock that reduces ballast settlement. Moreover, it is also clear that the geogrid is much more effective to eliminate the settlement of tracks built on the soft subgrade.

Ballast Breakage. Selected ballast aggregates were collected from beneath the sleepers as these locations appeared to be more prone to crushing due to higher load distribution. The ballast breakage index (BBI) was used to measure the breakage, and the results are presented in Table 10.1. The BBI is largest at the top of the ballast layer, and then reduced with depth. The BBI is less than 0.05 (BBI = 5%) which shows that minimal breakage occurs when obtaining these measurements. The higher stresses also cause more ballast breakage due to increased inter-particle contact forces. For the alluvial-deposit subgrade, there is a larger settlement due to more breakage occurs in these track sections. The smallest breakage occurs at the concrete-bridge decks, which shows that the shock mat has mitigated particle

Table 10.1 Measured ballast breakage

No	Subgrade conditions	Ballast breakage index (BBI) for		
		Top layer	Middle layer	Bottom layer
1	Alluvial-silty clay (track section A)	0.17	0.08	0.06
2	Concrete-bridge (track section B)	0.06	0.03	0.02
3	Siltstones (track section C)	0.21	0.11	0.09

degradation. The measured data clearly reveals how rubber shock mats installed on concrete bridge decks reduces ballast breakage. This is mainly because the rubber mat (shock mat) absorbs the kinetic energy caused by the moving train and as a result, less impact energy is being transmitted to ballast aggregates causing reduced breakage (Ngo et al. 2019).

10.5 Conclusions

This paper has introduced several innovative methods of using recycled rubber products such as under ballast mats (USM), under sleeper pats (USP), a synthetic energy absorbing layer for subballast (SEAL-a mixture of mining waste and rubber crumbs), rubber energy absorbing drainage sheets (READS), and a tyre cell confined capping layer. Large-scale cubical cyclic triaxial tests and drop hammer impact tests, and finite element modelling and field tests were carried out to investigate how efficiently these recycled rubber products could reduce ballast breakage and deformation, the salient findings are summarised as follows:

- Cubical cyclic triaxial tests revealed that the inclusion of USPs and UBMs and the use of the tyre cells to enhance the capping layer, efficiently reduced ballast degradation and track deformation (settlement and lateral dilation). Also, adding 10% rubber crumbs in SEAL significantly reduced ballast breakage and lateral displacement, and also ensured the settlement of the track specimen was comparable to a traditional track specimen.
- The drop hammer impact loading test showed that READS reduced ballast breakage and deformation (vertical and lateral) under impact loading regardless of the subgrade conditions (stiff or soft), albeit there was more reduction in settlement and BBI with a stiff subgrade.
- The FEM simulation of track reinforced with tyre cells further validated their ability to provide greater uniform lateral confining pressure to the track and thus reduce lateral movement.
- The field tests carried out on the sections of track with geogrids and shock mats showed enhanced deformation and reduced ballast degradation due to the geogrid-particle interlock and the energy absorbing property of the rubber mats. The geogrid reinforcement was also found to be more efficient on softer subgrades.

Acknowledgements The authors wish to acknowledge the financial support from the Australian Research Council Industrial Transformation Training Centre for Advanced Technologies in Rail Track Infrastructure (IC170100006) and Australian Research Council Discovery Project (DP180101916). The authors also wish to thank RM CRC, Global Synthetics Pty Ltd, and Foundation Specialists Group through Project R2.5.2. The efforts of previous PhD students and post-doctoral research fellows, Dr Nimbalkar, Dr Jayasuriya, Dr Navaratnarajah, Dr Qideng Sun, among others are also gratefully appreciate. Salient contents sourced from past articles (ASCE-J. Geotech. & Geoenviron. Engineering, Computers and Geotechnics, International J. Geomech., Journal of Materials in Civil Engineering, Transportation Geotechnics and Ground Improvement) have been reproduced here with modification and combination. The authors are also grateful to CME-technician at University of Wollongong for their assistance during the laboratory and field tests. The authors also wish to acknowledge the support from the Centre of Geomechanics and Railway Engineering (CGRE) at University of Wollongong during the above research.

References

Esmaeili M, Aela P, Hosseini A (2017) Experimental assessment of cyclic behavior of sand-fouled ballast mixed with tire derived aggregates. Soil Dyn Earthq Eng 98:1–11

Indraratna B, Lackenby J, Christie D (2005) Effect of confining pressure on the degradation of ballast under cyclic loading. Géotechnique 55(4):325–328

Indraratna B, Ngo NT, Rujikiatkamjorn C (2012b) Deformation of coal fouled ballast stabilized with geogrid under cyclic load. J Geotech Geoenviron Eng 139(8):1275–1289

Indraratna B, Nimbalkar S, Rujikiatkamjorn C (2014a) From theory to practice in track geomechanics-Australian perspective for synthetic inclusions. Transp Geotechn 1(4):171–187

Indraratna B, Nimbalkar S, Neville T (2014b) Performance assessment of reinforced ballasted rail track. Proc Inst Civil Eng-Ground Improv 167(1):24–34

Indraratna B, Nimbalkar SS, Ngo NT, Neville T (2016) Performance improvement of rail track substructure using artificial inclusions-Experimental and numerical studies. Transp Geotech 8:69–85

Indraratna B, Qi Y, Heitor A (2017a) Evaluating the properties of mixtures of steel furnace slag, coal wash, and rubber crumbs used as subballast. J Mater Civ Eng 30(1):04017251

Indraratna B, Sun Q, Heitor A, Grant J (2017b) Performance of rubber tire-confined capping layer under cyclic loading for railroad conditions. J Mater Civ Eng 30(3):06017021

Indraratna B, Sun Q, Grant J (2017c) Behaviour of subballast reinforced with used tyre and potential application in rail tracks. Transp Geotech 12:26–36

Indraratna B, Qi Y, Ngo TN, Rujikiatkamjorn C, Neville T, Ferreira FB, Shahkolahi A (2019) Use of geogrids and recycled rubber in railroad infrastructure for enhanced performance. Geosciences 9(1):30

Indraratna B, Nimbalkar S, Rujikiatkamjorn C (2012a) Future of Australian rail tracks capturing higher speeds with heavier freight

Indraratna B, Ferreira FB, Qi Y, Ngo TN (2018) Application of geo-inclusions for sustainable rail infrastructure under increased Axle loads and higher speeds. Innov Infrastr Solut 3(69)

Indraratna B, Qi Y, Tawk M, Heitor A, Rujikiatkamjorn C, Navaratnarajah SK (2020) Advances in ground improvement using waste materials for transportation infrastructure. Ground Improv. https://doi.org/10.1680/jgrim.20.00007

Jayasuriya C, Indraratna B, Ngo TN (2019) Experimental study to examine the role of under sleeper pads for improved performance of ballast under cyclic loading. Transp Geotech 19:61–73

Navaratnarajah SK, Indraratna B (2017) Use of rubber mats to improve the deformation and degradation behavior of rail ballast under cyclic loading. J Geotech Geoenviron Eng 143(6):04017015

Navaratnarajah SK, Indraratna B, Ngo NT (2018) Influence of under sleeper pads on ballast behavior under cyclic loading: experimental and numerical studies. J Geotech Geoenviron Eng 144(9):04018068

Ngo NT, Indraratna B, Ferreira FB, Rujikiatkamjorn C (2018) Improved performance of geosynthetics enhanced ballast: laboratory and numerical studies. Proc Inst Civil Eng-Ground Improv 171(4):202–222

Ngo TN, Indraratna B, Rujikiatkamjorn C (2019) Improved performance of ballasted tracks under impact loading by recycled rubber mats. Transp Geotech 20:100239

Nimbalkar S, Indraratna B (2016) Improved performance of ballasted rail track using geosynthetics and rubber shockmat. J Geotech Geoenviron Eng 142(8):04016031

Qi Y, Indraratna B, Heitor A, Vinod JS (2018a) Effect of rubber crumbs on the cyclic behavior of steel furnace slag and coal wash mixtures. J Geotech Geoenviron Eng 144(2):04017107

Qi Y, Indraratna B, Vinod JS (2018b) Dynamic properties of mixtures of waste materials. In: GeoShanghai international conference. Springer

Qi Y, Indraratna B, Vinod JS (2018c) Behavior of steel furnace slag, coal wash, and rubber crumb mixtures with special relevance to stress-dilatancy relation. J Mater Civ Eng 30(11):04018276

Qi Y, Indraratna B, Heitor A, Vinod JS (2018d) Closure to "effect of rubber crumbs on the cyclic behavior of steel furnace slag and coal wash mixtures" by Qi Y, Indraratna B, Heitor A, Vinod JS (eds) J Geotechn Geoenviron Eng 145(1):07018035

Qi Y, Indraratna B, Heitor A, Vinod J (2019a) The influence of rubber crumbs on the energy absorbing property of waste mixtures. In: Sundaram SJR, Havanagi V (eds) Geotechnics for transportation infrastructure. Springer, Singapore, pp 271–281

Qi Y, Indraratna B, Coop MR (2019b) Predicted behavior of saturated granular waste blended with rubber crumbs. Int J Geomech 19(8):04019079

Qi Y, Indraratna B (2020) Energy-based approach to assess the performance of a granular matrix consisting of recycled rubber, steel furnace slag and coal wash. J Mater Civil Eng, (ASCE) 32(7):04020169. https://doi.org/10.1061/(ASCE)17MT1943-5533.0003239

Signes CH, Fernández PM, Perallón EM, Franco RI (2015) Characterisation of an unbound granular mixture with waste tyre rubber for subballast layers. Mater Struct 48(12):3847–3861

Sol-Sánchez M, Moreno-Navarro F, Rubio-Gámez MC (2015) The use of elastic elements in railway tracks: a state of the art review. Constr Build Mater 75:293–305

Chapter 11
Life Cycle Sustainability Assessment: A Tool for Civil Engineering Research Prioritization and Project Decision Making

Alena J. Raymond, Jason T. DeJong, and Alissa Kendall

11.1 Introduction

In 2017, the buildings and construction sector accounted for 36% of total energy use and 39% of carbon dioxide emissions, globally (UN Environment and International Energy Agency 2017). These estimates will continue to increase due to population growth and urbanization, among other factors. Given the serious implications of these projections, civil engineers in academia and industry have an opportunity and responsibility to contribute solutions for sustainable development that will reduce the impacts of civil infrastructure projects. While researchers around the world are working to develop more sustainable strategies for infrastructure construction, few are asking the questions: "What are the environmental impacts of current methods?", "Where are there opportunities to reduce impacts and improve sustainability?", and "How do I know that my proposed solution is, in fact, more sustainable than existing methods?". In this research, an integrated life cycle sustainability assessment (LCSA) approach was developed to help answer these questions and drive sustainability-oriented research, development, and deployment of emerging technologies in industry.

The proposed LCSA methodology is designed to provide a more holistic approach compared to current project decision making, which aims to minimize cost while satisfying safety and performance criteria (Holt et al. 2010). Instead, LCSA quantitatively evaluates the environmental, economic, and social impacts and/or benefits of a product or system over its entire life cycle (i.e., from "cradle to grave"). LCSA also provides a framework for weighing the tradeoffs of alternative products or systems, while avoiding shifting burdens (e.g., between life cycle stages, regions, stakeholders, generations, or types of impacts).

A. J. Raymond (✉) · J. T. DeJong · A. Kendall
University of California, Davis, CA 95616, USA
e-mail: ajraymond@ucdavis.edu

This paper describes the developed LCSA tool for research evaluation and advancement. An example focusing on liquefaction mitigation strategies is given to demonstrate how LCSA can be applied to existing and incipient technologies to make comparisons and guide research and development efforts toward more sustainable solutions.

11.2 LCSA Methodology

11.2.1 Background on LCSA

Typically, LCSA integrates environmental life cycle assessment, life cycle costing, and social life cycle assessment to evaluate impacts to the three pillars, or triple bottom line, of sustainability: environmental protection, economic growth, and social progress. Environmental life cycle assessment (LCA) characterizes, quantifies, and interprets a product or system's environmental impacts along the entire supply chain, from extraction of raw materials to end of life. Life cycle cost analysis (LCCA) and social life cycle assessment (S-LCA) complement environmental LCA by evaluating the relevant costs and social impacts associated with the product or system that are incurred by stakeholders over the equivalent life cycle defined in the LCA. LCA methods have been standardized in the International Organization for Standardization (ISO) standards ISO 14040:2006 and ISO 14044:2006 (ISO 2006; ISO 2006). Similarly, many standards and guidelines exist for LCCA of civil infrastructure systems (ASTM International 2017; Highway and Administration: Life-cycle cost analysis in pavement design 1998; ISO 2017). While guidelines for S-LCA of products also exist (UNEP 2009), these methods are less developed and not as widely used (Neugebauer et al. 2015). In place of a formal S-LCA, the developed LCSA tool estimates the social damage costs associated with greenhouse gas (GHG) emissions (e.g., carbon dioxide, methane, and nitrous oxide) using the methods and models employed by the United States Environmental Protection Agency (Interagency Working Group on Social Cost of Greenhouse Gases: Technical update of the social cost of carbon for regulatory impact analysis under executive order 2016). While not comprehensive, the social cost of GHG emissions serves as an indicator of the potential socioeconomic impacts of existing and emerging technologies.

11.2.2 Important Terminology

Knowledge of the following terminology is critical to operationalize the LCSA methodology as a tool for research prioritization and project decision making:

- **System Boundary**: The system boundary describes the life cycle stages and unit processes included in and excluded from the scope of a LCSA and is

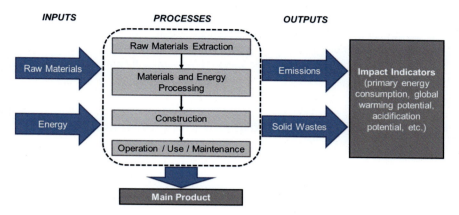

Fig. 11.1 Example process flow diagram

often accompanied by a process flow diagram of the studied product or system (Fig. 11.1).

- **Functional Unit**: The functional unit describes and quantifies the identified function(s) and performance characteristics of the studied product or system (ISO 2006). The functional unit serves as a reference unit for the LCSA results and a basis for comparison of alternative products or systems, making its definition particularly important when comparing emerging technologies against conventional or business-as-usual (BAU) methods used in industry.
- **Inventory Data**: A life cycle inventory (LCI) catalogs the relevant environmental and economic flows associated with the product or system over its defined life cycle. For example, environmental flows include inputs such as energy and raw materials as well as outputs such as air emissions, water pollutants, and solid wastes. The LCI may consist of primary data based on actual processes and their operating conditions as well as background data from reference LCI databases, which typically represent average industry processes.
- **Indicators**: Through use of characterization factors, which have been proposed by various life cycle impact assessment (LCIA) methodologies, the LCI results are translated into indicators that represent potential impacts to areas of protection (e.g., human health, natural environment, and natural resources). The most well-known and widely adopted environmental indicator is the Intergovernmental Panel on Climate Change's global warming potential (GWP), which converts GHG emissions into units of carbon dioxide equivalents (CO_2 eq.) (Myhre et al. 2013).
- **Hotspots**: A key step in the interpretation of LCSA results is the identification of environmental and economic hotspots, or life cycle stages, processes, and flows with large contributions to one or several impact indicators (Laurent et al. 2020). Hotspot analysis highlights significant areas of concerns and helps to identify research tasks that should be prioritized to improve the sustainability of emerging technologies.

- **Benchmarks**: Sustainability benchmarks, or thresholds, represent the impacts of conventional technologies or routine practices in industry. These benchmarks are needed to make transparent and objective comparisons between new technologies and BAU methods and to move the standard of practice toward more sustainable solutions.

11.2.3 LCSA for Research Evaluation and Advancement

LCSA can have an important role in the research and development of new civil engineering technologies (Arvidsson et al. 2018). In the developed LCSA approach, LCSA is directly integrated into the research and development process to:

- identify concerns regarding the proposed technology's environmental impacts, feasibility, cost, or scale of transformation (i.e., the existing or future market for the technology);
- provide sustainability-oriented recommendations for future research priorities and continued development of the proposed technology;
- describe the proposed technology's potential for sustainability gains (e.g., environmental impact reductions or cost savings) relative to existing BAU methods; and
- help researchers become more sustainability-minded about their work overall.

The developed LCSA tool is iterative and flexible to maximize its utility across all stages of research and development, from small scale feasibility testing to field-scale demonstrations (Fig. 11.2). A preliminary and largely qualitative LCSA is performed to evaluate feasibility and identify environmental and/or economic hotspots early in the development of a new technology when limited quantitative data are available. In later stages of technological development, the LCSA becomes increasingly quantitative as data become available (e.g., about material quantities and equipment required for construction). As the technology nears deployment in industry, a comprehensive, quantitative LCSA is performed to compare the proposed technology to sustainability

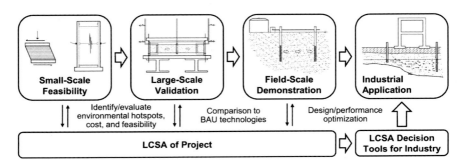

Fig. 11.2 LCSA framework for research evaluation and advancement

benchmarks of BAU methods. Ultimately, the integrated LCSA approach can help advance the research and development of competitive and sustainable technologies for infrastructure construction.

11.3 LCSA Example

Mitigation of soil liquefaction during earthquakes is a geotechnical engineering challenge with both current and emerging solutions. Several BAU ground improvement methods exist (e.g., deep soil mixing, deep dynamic compaction, and compaction grouting). However, many use large quantities of Portland cement and mechanical energy to improve the geotechnical properties of soils.

Microbially induced calcite precipitation (MICP) is a bio-mediated process being leveraged by researchers as a new ground improvement technology for liquefaction mitigation, among other geotechnical and geoenvironmental applications (Gomez et al. 2017; San Pablo et al. 2020). The MICP technology uses natural microbial enzymatic activity to precipitate calcium carbonate, or calcite, on soil particle surfaces and at particle contacts, thus improving the stiffness and shear strength of granular soils. The MICP reaction chemistry is further described in the literature (Gomez et al. 2018; Lee et al. 2019).

MICP offers possible reductions of environmental impacts compared to BAU methods for liquefaction mitigation because it does not require Portland cement, it only requires pumping energy during construction, and it provides a nondestructive method for improving soils under existing infrastructure. To evaluate this hypothesis, a LCSA of BAU methods was performed to benchmark the environmental impacts of conventional improvement techniques. To estimate the potential impacts of MICP, LCSA was integrated into MICP research and development to provided sustainability-oriented recommendations for future research priorities and continued development of the technology. When implemented in research projects, LCSA becomes an iterative process that evolves over time as projects mature through different stages of development. Thus, while the LCSA of MICP is ongoing, sustainability improvements have already been implemented (e.g., through identification of environmental hotspots), and areas for future research and technology advancement have been identified.

11.3.1 MICP Formulation Optimization at the Laboratory Scale

Based on the results of MICP feasibility testing in the laboratory, a preliminary LCSA of MICP was performed to evaluate the impacts of materials production and use per 1 m^3 of treated soil. This initial assessment identified urea and calcium chloride as

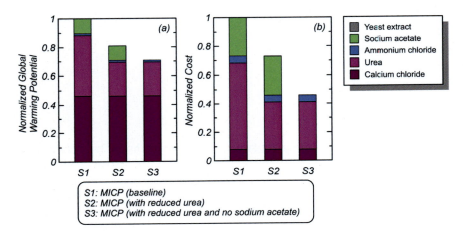

Fig. 11.3 Potential reductions in **a** environmental impact and **b** reagent cost realized by preliminary LCSA results and subsequent MICP research

environmental and economic hotspots and found that the ammonium (NH_4^+) by-products of the MICP reaction, which stem from urea use, have potentially large environmental impacts.

These findings highlighted the need for further research focused on optimizing the MICP formulation. In response, MICP researchers performed laboratory testing to investigate the potential to reduce urea in the MICP formulation. Reductions in urea consumption reduce NH_4^+ by-products and minimize associated environmental impacts and treatment costs. The experimental results demonstrated that urea use could be reduced by approximately 45% without compromising the geotechnical performance MICP-treated soils. This reduction in urea translated to reductions in total GWP and cost by 19% and 27%, respectively (Fig. 11.3).

11.3.2 Evaluation of Meter-Scale MICP Tests

Using the experimental results of meter-scale column tests (San Pablo et al. 2020; Lee et al. 2019), a more comprehensive LCSA was performed to compare the impacts and costs of different MICP treatment approaches, namely augmentation (i.e., injection of non-native bacteria) and stimulation (i.e., enrichment of indigenous microorganisms). This LCSA expanded on the preliminary study to reflect reduced urea use in the MICP formulation and include impacts associated with pumping (i.e., during construction operations). The study also evaluated the potential environmental benefits of NH_4^+ by-product removal using a rinsing technique. Figure 11.4 compares the impacts of each treatment approach both with and without rinsing to remove the

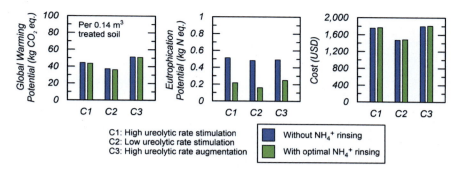

Fig. 11.4 LCSA results of different MICP treatment approaches with and without ammonium removal (Target improvement: Vs = 150 m/s)

NH_4^+ by-products from the soil pore space. From the results of the experimental work coupled with LCSA, several observations were made:

- Stimulation is potentially more sustainable than augmentation, and the use of lower ureolytic rates can further improve sustainability by minimizing reactions during injections, resulting in greater spatial uniformity and extent of bio-cementation compared to other approaches.
- Rinsing can successfully remove NH_4^+ by-products; however, there is a tradeoff between the environmental benefits of NH_4^+ by-product removal and the life cycle impacts and costs associated with the rinsing technique. These impacts and benefits can be balanced using an optimal rinsing scheme.
- Optimal rinsing significantly reduces eutrophication impacts while minimally affecting cost and other environmental indicators like GWP.
- With further development focused on prevention and management of NH_4^+ by-products, MICP could become more viable and competitive relative to BAU methods for liquefaction mitigation.

11.3.3 Development of BAU Benchmarks for Comparison to MICP

In a parallel study and through collaboration with geotechnical consultants and contractors, a LCSA of BAU methods was performed to develop realistic, industry-representative benchmarks of the environmental and economic impacts of conventional ground improvement techniques used in practice for liquefaction mitigation (Fig. 11.5). To quantify the impacts of each method, typical scopes of work were established with guidance from specialty contractors that included high (i.e., worst case) and low (i.e., best case) bounds to reflect the risk and uncertainty inherent to ground improvement design and construction activities. LCSA results showed that deep soil mixing and compaction grouting are the most impactful and costly methods

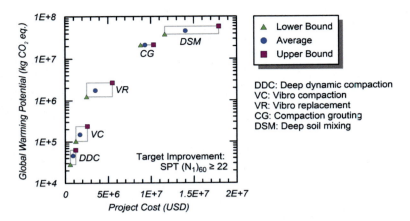

Fig. 11.5 LCSA results of BAU ground improvement methods for liquefaction mitigation (per 248,000 m³ of improved soil)

due to the use of Portland cement. Vibratory methods and deep dynamic compaction are more sustainable, despite that their impacts and costs are more variable and uncertain.

Future work will upscale the LCSA of MICP from the meter-scale to field-scale in order to directly compare the impacts of MICP against those of BAU methods for liquefaction mitigation. The scope will include the evaluation of relevant impacts and costs that would be incurred during field-scale implementation of MICP, including materials transportation, equipment mobilization and use, construction consumables, waste management, and labor costs. The study may identify particular applications where MICP offers sustainability gains over BAU methods (e.g., liquefaction mitigation under existing structures where many BAU methods are not feasible or would be very expensive). However, this type of advanced comparative LCSA requires that the considered designs for MICP and the BAU methods deliver the same functionalities and performance characteristics. Thus, further experimental research is needed to better understand and relate different improvement mechanisms (e.g., biocementation versus densification) to the systems-level response of liquefiable soils during earthquakes, which typically manifests as differential settlements or lateral displacements.

11.4 Conclusions

This paper describes the developed LCSA tool for research prioritization and project decision making, which serves to inform sustainability-oriented research, development, and deployment of emerging technologies. The integrated LCSA approach is adaptable and can be applied to technologies at different stages of development and with different applications. The utility of the LCSA tool developed by this research has been demonstrated and improved through its application to a diverse portfolio of projects at different stages of development and with different applications. In addition to ground improvement methods, the LCSA framework has been applied to dust mitigation and erosion control techniques (Raymond et al. 2020) and geotechnical site characterization tools (Purdy et al. 2020), among others. Understanding the impacts of BAU methods through LCSA can inform project planning and decision making in industry, and it provides a baseline for researchers against which new technologies can be compared. LCSAs of incipient technologies are critical for demonstrating the sustainability and cost effectiveness of new solutions relative to BAU methods to potential users and can help encourage early adoption of more sustainable solutions by practicing engineers. While traditional civil engineering education and practice has been slow to adopt the use of life cycle thinking and quantitative environmental impact assessment methods, the developed LCSA approach could be implemented in academia or industry to identify opportunities to reduce impacts and improve sustainability in the design and construction of engineered systems.

Acknowledgements This material is based upon work supported by the Engineering Research Center Program of the National Science Foundation (NSF) under NSF Cooperative Agreement EEC-1449501. Any opinions, findings, and conclusions or recommendations expressed in this material are those of the authors and do not necessarily reflect those of the NSF.

References

Arvidsson R, Tillman A, Sandén BA, Janssen M, Nordelöf A, Kushnir D, Molander S (2018) Environmental assessment of emerging technologies: recommendations for prospective LCA. J Ind Ecol 22:1286–1294

ASTM International (2017) ASTM E917-17 standard practice for measuring life-cycle costs of buildings and building systems. ASTM International, West Conshohocken, PA

Federal Highway Administration: Life-cycle cost analysis in pavement design. FHWA-SA-98-079. Pavement Division Interim Technical Bulletin

Gomez MG, Anderson CM, Graddy CMR, DeJong JT, Nelson DC, Ginn TR (2017) Large-scale comparison of bioaugmentation and biostimulation approaches for biocementation of sands. J Geotech Geoenv Eng 143(5):04016124

Gomez MG, Graddy CMR, DeJong JT, Nelson DC, Tsesarsky M (2018) Stimulation of native microorganisms for biocementation in samples recovered from field-scale treatment depths. J Geotech Geoenviorn Eng 144(1):04017098

Holt DGA, Jefferson I, Braithwaite PA, Chapman DN (2010) Sustainable geotechnical design. In: Proceedings of the GeoFlorida 2010. ASCE, Reston, VA, pp 2925–2932

Interagency Working Group on Social Cost of Greenhouse Gases: technical update of the social cost of carbon for regulatory impact analysis under executive order 12866 (2016)

ISO (International Organization for Standardization) (2006) ISO 14040:2006 Environmental management—Life cycle assessment—Principles and framework. Geneva, Switzerland

ISO (2006) ISO 14044:2006: environmental management—Life cycle assessment—Requirements and guidelines. Geneva, Switzerland

ISO (2017) ISO 15686-5:2017: buildings and constructed assets—Service life planning—Part 5: life-cycle costing. Geneva, Switzerland

Laurent A, Weidema BP, Bare J, Liao X, Maia de Souza D, Pizzol M, Sala S, Schreiber H, Thonemann N, Verones F (2020) Methodological review and detailed guidance for the life cycle interpretation phase. J Ind Ecol 13012

Lee M, Gomez MG, San Pablo ACM, Kolbus CM, Graddy CMR, DeJong JT, Nelson DC (2019) Investigating ammonium by-product removal for ureolytic bio-cementation using meter-scale experiments. Sci Rep 9:18313

Myhre G, Shindell D, Bréon FM, Collins W, Fuglestvedt J, Huang J, Koch D, Lamarque JF, Lee D, Mendoza B, Nakajima T, Robock A, Stephens G, Takemura T, Zhang H (2013) Anthropogenic and natural radiative forcing. In: Climate change 2013: the physical science basis. Contribution of Working Group I to the Fifth Assessment Report of the Intergovernmental Panel on Climate Change. Cambridge University Press, Cambridge, UK and New York, NY, USA

Neugebauer S, Martinez-Blanco J, Scheumann R, Finkbeiner M (2015) Enhancing the practical implementation of life cycle sustainability assessment—Proposal of a tiered approach. J Clean Prod 102(2015):165–176

Purdy CM, Raymond AJ, DeJong JT, Kendall A, Krage C, Sharp J (2020) Life cycle sustainability assessment of geotechnical site investigation. Can Geotech J

Raymond AJ, Kendall A, DeJong JT, Kavazanjian E, Woolley MA, Martin KK (2020) Life cycle sustainability assessment of fugitive dust control methods. J Constr Eng Manag

San Pablo ACM, Lee M, Graddy CMR, Kolbus CM, Khan M, Zamani A, Martin N, Acuff C, DeJong JT, Gomez MG, Nelson DC (2020) Meter-scale biocementation experiments to advance process control and reduce impacts: examining spatial control, ammonium by-product removal, and chemical reductions. J Geotech Geoenviorn Eng 146(11):04020125

UN Environment and International Energy Agency: towards a zero-emission, efficient, and resilient buildings and construction sector. Global Status Report 2017

UNEP (2009) Guidelines for social life cycle assessment of products

Chapter 12
Role of the Indonesian Society for Geotechnical Engineering in the Development of Sustainable Earthquake-Resilience Infrastructure in the Recent Years

Arifan Jaya Syahbana, Masyhur Irsyam, Delfebriyadi Delfebriyadi, Mahdi Ibrahim Tanjung, Rena Misliniyati, Mohamad Ridwan, Fahmi Aldiamar, Nuraini Rahma Hanifa, Arifin Beddu, and Agus Himawan

12.1 Introduction

Indonesia is a country located in the Southeast Asia region. Due to tectonic conditions that are also unique, which is located between the Indo Australian plate, the Asian Plate and the Philippine Plate, frequent earthquakes are caused by the plate's movement (Altamimi et al. 2016). In the past 15 years, there have been many large

---

A. J. Syahbana (✉) · M. Irsyam · D. Delfebriyadi · M. I. Tanjung · R. Misliniyati · F. Aldiamar · A. Himawan
Faculty of Civil and Environmental Engineering, Bandung Institute of Technology, Bandung 40132, Indonesia
e-mail: arifanjaya@s.itb.ac.id

A. J. Syahbana · M. Irsyam · D. Delfebriyadi · M. I. Tanjung · R. Misliniyati · M. Ridwan · N. R. Hanifa
Research Center for Disaster Mitigation, Bandung Institute of Technology, Bandung 40132, Indonesia

M. Ridwan
Research Institute for Housing and Human Settlements, Ministry of Public Works and Housing, Bandung 40393, Indonesia

F. Aldiamar
Institute of Road Engineering, Ministry of Public Works and Housing, Bandung 40216, Indonesia

N. R. Hanifa
Center for Earthquake Science and Technology, Bandung Institute of Technology, Bandung 40132, Indonesia

A. Beddu
Faculty of Civil Engineering, Tadulako University, Palu 94148, Indonesia

© The Author(s), under exclusive license to Springer Nature Singapore Pte Ltd. 2023
H. Hazarika et al. (eds.), *Sustainable Geo-Technologies for Climate Change Adaptation*, Springer Transactions in Civil and Environmental Engineering, https://doi.org/10.1007/978-981-19-4074-3_12

earthquakes and secondary disasters that have followed. This phenomenon resulted in many materials and non-material losses, such as more than 140,000 fatalities and damage to infrastructures such as buildings, roads, bridges, and residential homes (BNPB 2019). Responding to this phenomenal event, the government, through the Ministry of Public Works and Housing supported by the Indonesian Society for Geotechnical Engineering (ISGE) conducted several activities to reduce the impact of the natural disaster. In this paper, several things related to earthquake hazard adaptation in Indonesia will be presented, including the introduction of Indonesian geotechnical standards, updating of the Indonesian Earthquake Map (including earthquake source relocation, earthquake source updating, GMPE studies), studies of earthquake events and liquefaction in Palu, establishment Nalodo Research Center, earthquake hazard and risk studies in Jakarta and the preparation of bridge design rules. Those things are described in the below section.

12.2 Sedimentary Basin Effect in Seismic Hazard Map Design of Jakarta

Understanding the basin effect can control the characteristics of ground motion at soil surface is very important in seismic hazard analysis, especially the amplitude of acceleration for a long period. Some studies indicate that the bedrock depth of Jakarta varies from 300 to 600 m. It means that the geometry of subsoil structure of Jakarta is irregular configuration. The parametric study based on this condition was performed in the 1D calculation on two cohesive soil deposit profiles, medium clay (site SD) and soft clay (site SE). To observe its impact on the surface response spectra, the bedrock layers of both profiles were located at any variation of depths of 300, 400, 500, and 600 m.

Figure 12.1 presents the effect of depth of bedrock to spectral acceleration at ground surface. The bedrock depth of 300 m on both site profiles (site S_D and site S_E)

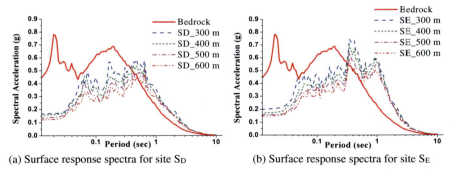

Fig. 12.1 The effect of depth of bedrock on the surface response spectra for a megathrust earthquake with a return period of 2500 years scaled at a spectral period of 0.0 s

yields the largest response of spectral accelerations among other depth variations. The surface spectral accelerations tend to decrease as the depth of the bedrock increases. Deeper bedrock elevation causes the distance of the seismic wave motion propagates longer to soil surface. Thus, the dissipated energy absorbed by the soil deposit also increased. The results show that amplitude of seismic wave motion at ground surface tends to decrease. At long periods the effect of bedrock depth was insignificant. The results also show that at short period the amplitude of spectral acceleration tends to be larger for the shallower bedrock depth and vice versa.

12.3 Jakarta City Risk Assessment

Jakarta city is one of the densely populated metropolises located in seismic prone regions. A seismic risk assessment has been made for the residential function zones. The seismic risk to buildings was quantified by assessing the probability of the buildings being in damage states, and the seismic risk to the population was quantified by estimating the numbers of casualties resulting from damage to the buildings.

Masonry building and concrete frame with unreinforced masonry infill walls (infilled frame) are assumed to be the dominant building constructions in the residential function zones. The percentage distribution of the building typology for each zone area is assumed by using the zones information in Fig. 12.2. The total number of buildings in each geounit and its distribution among the typologies is computed. Hazard assessment for Jakarta city in this study represents the earthquake scenario $M_w = 7$ and $R = 117$ km subjected to the intraslab subduction mechanism. The site

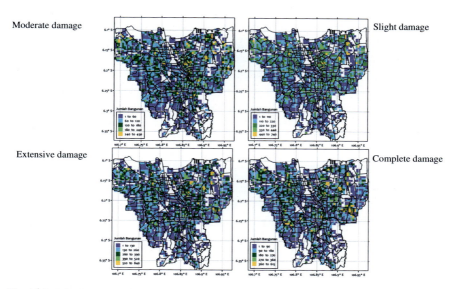

Fig. 12.2 Distribution of damage building of confined masonry due to earthquake scenario (plotted in 0.5×0.5 km^2 geounit)

Table 12.1 Projected Casualties in DKI Jakarta 2030 due to Earthquake Shaking based on Earthquake Scenario $M_W = 7.0$; R = 117 km. (Severity in person unit)

Intensity (MMI)	Estimated population exposure	Minor injured Severity 1	Hospitalized injured Severity 2	Severity 3	Fatality Severity 4
VI	97,017	381	93	15	30
VII	11,734,805	151,325	45,091	8,080	16,038
VIII	42,088	1,490	470	86	171
	Total	153,196	53,834		16,239

response analyses are carried out to estimate the peak ground acceleration (*PGA*) at the ground surface.

The fragility curve would be a key component in seismic damage assessment. For this study, the standard fragility curves are then adopted from HAZUS-MH MR4 (Federal Emergency Management Agency (FEMA) 2003). The three structural systems classify into the building of low-rise concrete frame with unreinforced masonry infill walls type (C3L) for low code to special low code seismic design level and building of low-rise unreinforced masonry bearing walls type (URML) for low code seismic design level. As a result, the distribution of damage building (e.g., confined masonry) for earthquake scenario applied to population projected in the year 2030 are shown in Fig. 8. provides a summary of damage building of the particular category at each damage state corresponding to the Benioff mechanism ($M_w = 7.0$; $R = 117$ km). The number of buildings at complete damage state for all types of buildings is projected to be 318,416 (15.2%).

The estimates of physical damages are converted to estimates of casualties using human injury and death rates applied from *HAZUS-MH MR4*. The fatalities and the injuries resulting from the scenario earthquake would be principally attributable to the failure of buildings. Four casualty levels were classified into namely severity 1 (requiring basic medical aid), severity 2 (requiring hospitalization), severity 3 (life-threatening and requiring immediate attention), and severity 4 (fatalities). Those casualties were related to given damage states between no collapsed and collapsed off the building. In this study, the casualty was estimated at nighttime of the day as the worst case because it considered that the residential occupancy load was maximum. These casualty rates are applied to the total number of affected persons. Table 12.1 provides a summary of the casualties estimated for this earthquake scenario. The occupancy of the building is based on the nighttime as if the earthquake occurred while people were at home.

12.4 Palu Earthquake Analysis and Report 2018

The 2018 Palu earthquake triggered a massive, far-reaching flow slide, which caused many fatalities. Flow slide as vast and so far as a result of an unprecedented earthquake in the modern world. In the city of Niigata, the land which is reacted has a slope of approximately 1% with a maximum lateral movement of 4 m. This phenomenon is very significant compared to the movements that occur in Palu City. Petobo ±2% slope with ±1100 m movement, Jono Oge has ±2.8% slope with ±1850 m movement, ±1.8% slope with ±193 m movement, Sibalaya slope ±4.2% with ±588 m movement, and Balaroa slope ±4.9% with a movement of ±400 m. The phenomenon of deformation of this magnitude is not yet clearly understood how the actual mechanism of events, large deformations due to earthquakes would be challenging to explain without the mechanism of pore redistribution in the sand layer (Kokusho and Fujita 2002). This phenomenon is rare, but if it occurs in a dense settlement, it will have enormous consequences. Understanding the flow slide phenomena that occur in Palu is very valuable, especially in mitigating the flow slide disaster due to the next earthquake.

Trenching in Lolu is carried out on the boundary of a movable and immovable area, while in Sibalaya, the test pit is carried out in the middle of the flow slide area. In both locations, relatively have the same layer pattern that is alternating sand or gravelly sand with silt. The potential layer is the dominant layer of gradation of loose sand, and when the earthquake is below the groundwater level (Fig. 12.3).

Figure 12.4 shows one of the results of the cyclic test with a constant volume of Lolu and Sibalaya sand using a simple shear tool that illustrates the cyclic response of pore pressure and shear strain to the cyclic amount. Liquefaction conditions are considered to occur in response to pore pressure (ru) = 1, or double amplitude shear strain (γDA) = 7.5%.

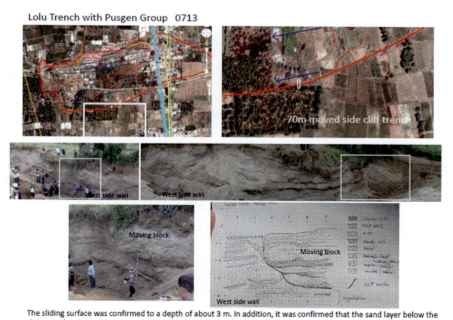

The sliding surface was confirmed to a depth of about 3 m. In addition, it was confirmed that the sand layer below the sliding surface had no lamina and was in a loose and easily liquefied Layer.

Fig. 12.3 Trenching activities and layers on the Lolu Trenching Wall

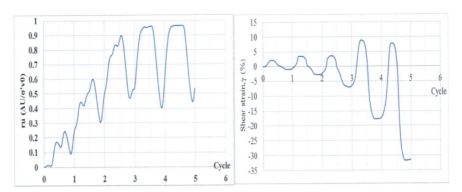

Fig. 12.4 Lolu Sand response ($Dr = 42\%$, CSR = 0.26, $\sigma'v0 = 100$ kPa) Due to Cyclic Load, **a** pore water pressure response, **b** shear strain response

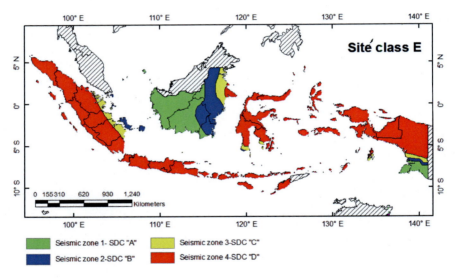

Fig. 12.5 Example of proposed seismic performance zones using the value of SD1 for site class E (Aldiamar et al. 2013)

12.5 Development and Updating of Standard on Seismic Load Design for Conventional Bridges in Indonesia

In 1973, the Indonesian government, with the assistance of New Zealand consultant, did the seismic study for building/structure to obtain seismic zone map, standard, and regulations on seismic resistance design for building/structure in Indonesia, namely Standar Perencanaan Ketahanan Gempa untuk Jembatan Jalan Raya (SNI 03-2833-1992) and Standar Perencanaan Ketahanan Gempa untuk Jembatan (SNI 2833:2008).

Another updating on seismic design standard was made in 2013 by adopting the AASHTO (2012) design criteria for a 1000-year return period (7% probability of exceedance in 75 years). This was a significant change, considering the previous national standard set a 500 year return period (10% probability of exceedance in 50 years). The seismic hazard map was created using the methodology and seismic source parameters established by Tim Revisi Peta Gempa (2010) and had been publicized in seminars at Puslitbang Jalan dan Jembatan (Asrurifak et al. 2012) and 18th Southeast Asian Geotechnical Conference.

Bridge damage due to earthquake could be severe and moderate. Lack of knowledge in designing seismic load for the bridge could increase the risk of bridge conditions under large earthquake events. Continuous updates on the seismic source and seismic hazard standard could reduce damage to the bridge and ensure the mobility to the earthquake-affected area after a disaster due to earthquakes happen. Other changes in the latest national standard were site class and site factors determination and design response spectrum calculation. Each bridge shall be assigned to one of the four seismic zones. The latest updates for bridge design standards are currently conducted using the 2017 seismic hazard map by the National Earthquake Center of Indonesia under the Ministry of Public Works.

12.6 Indonesian Earthquake Hazard Map 2017 Development

Indonesia Earthquake Hazard Map 2017 is an improvement product from PusGen to revise the previous one Indonesia Earthquake Hazard Map 2010 adopted by SNI 1726:2012. In this point, there are some significant differences between the previous seismotectonic models (used in SNI 1726-2012) and those that are updated (Irsyam et al. 2017), namely: (1) Revision of the Sumatra Fault Zone (SFZ) segment. However, the new SFZ seismotectonic model is similar to that used in the SNI-2012 PSHA, Toru-Renun-Tripa, Aceh-Seulimeum, Kumering, and other fault segments near the Sunda Strait have been revised with more appropriate locations (Natawidjaja 2018). SFZ now has an overall slip rate of between 10 and 14 mm/year, with no downward trend seen from north to south that was used for the 2010 map (Natawidjaja et al. 2017). (2) Reconsider the Sumatra Megathrust Zone (SMZ). Significant changes for the SMZ were an increase in Mmax for the Mentawai segment from 8.5 to 8.9, for the southern segment (Enggano) from 8.2 to 8.4, and the addition of the Sunda Strait segment with Mmax 8.7. (3) The crustal fault on the island of Java was restructured into a more accurate type of fault in the Cimandiri Fault segment, and added several additional faults around the Baribis Fault, Semarang Push, and Kendeng-Folding Belt and Push Rembang. The parameters for the Lembang fault are updated from the results of a detailed study of the Lembang Fault (Daryono et al. 2019). Finally (4) Adjustment of crustal geometry of crust and slip rate in Sulawesi and Papua.

The data used in the 2017 Indonesia Earthquake Disaster Map are compiled from the history of earthquake events with Magnitude $M_w > 4.5$ that occurred in and around Indonesia from 1900 to 2016 combined from many sources such as (a) International Seismology Center (ISC), (b) US National Geological Survey Earthquake Information Center (NEIC-USGS), (c) earthquake catalog that has been relocated by Engdahl, and (d) catalog of the BMKG hypocenter. Some of the sources of the earthquake showed that the total number of earthquakes that occurred in Indonesian territory between 1900 and 2016 with a magnitude of $M_w > 4.5$ was 51,855. These data are then labeled Catalog of the Indonesian National Earthquake Study Center (PuSGeN). On the other hand, since there are currently no Ground Motion Prediction Equations (GMPEs) developed specifically for the Indonesian region, we use GMPEs derived in other regions that have similar tectonic environments.

The Team for Updating of Seismic Hazard Maps of Indonesia 2017 has delivered national maps of peak ground acceleration (PGA) and spectral accelerations for periods of 0.2 and 1.0 s at bedrock from deterministic and probabilistic seismic hazard analyses. The maps represent eight levels of a hazard: 20, 10, and 5% probability of exceedance in 10 years; 7% probability of exceedance in 75 years; 10 and 2% probability of exceedance in 50 years, and 2 and 1% probability of exceedance in 100 years. The maps have been signed by the Minister of Public Works and People Housing of the Republic of Indonesia in September 2017. They have become the official national hazard maps of Indonesia (Fig. 12.6).

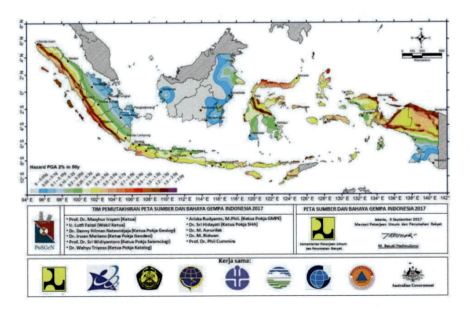

Fig. 12.6 Example of bedrock PGA in Indonesia for reoccurrence 2475 years, i.e., 2% exceeded for 50 years (PusGen 2017)

12.7 Establishment of Nalodo Research Center

The primary mission of establishing the Nalodo Research Center is to massively improve Nalodo's research by encouraging scientists to carry out sophisticated and innovative research methods including the development and dissemination of the latest technological systems, encouraging high-quality Nalodo research throughout the world through the collaboration of researchers, institutions, associations, and industry that effectively promotes Nalodo's research. Therefore, Tadulako University, as a State Higher Education institution in Central Sulawesi Province, took the initiative to compile and propose the establishment and management plan of the Nalodo Research Center. A series of Nalodo Research Center programs and activities are described in the Road Map scheme, which consists of Short-term (2019–2021) which focuses on the capacity building of research institutions with an emphasis on infrastructure, the fulfillment of laboratory testing equipment, subsurface survey tools, and increase in human resource capacity. Medium-term (2021–2025): programs for institutional capacity building will be more directed to be able to produce hazard and risk mapping applications with high-resolution data at the microzonation level and also at the performance levels, namely strengthening the social side and risk communication to stakeholders. Long-term (2025–2030): NRC development in Nalodo Disaster Mitigation will focus more on global services to Nalodo's research needs and make NRC an icon of Indonesia's Nalodo disaster mitigation research at the level of forming a knowledge base and technological innovation.

12.8 Conclusions

This paper has delivered some lessons that can be drawn as follow: from several large earthquakes that have repeatedly occurred in Indonesia, including the phenomenal 2018 earthquake in Palu City, which has increased public and government awareness about seismic activities in Indonesia. Over the last fifteen years, many research institutes, universities, including professional associations such as the Indonesian Society for Geotechnical Engineering (ISGE), have carried out significant studies to understand the danger of earthquakes and their contribution to reducing the impact of significant earthquakes in Indonesia in the future. Some of the most recent efforts in Indonesia to mitigate the impact of earthquake hazards are briefly described in this paper. These actions included updating the Indonesian seismic hazard map from 2010 to 2017, revising and updating the bridge design code, liquefaction studies, and establishing Nalodo Research Center, studies of earthquake hazard and risk also basin effects in Jakarta. Furthermore, ISGE is also creating Geotechnical Indonesian National Standard of 2017 and still in the process of preparing ground motion standard codes to be applied nationally in Indonesia.

References

Aldiamar F, Irsyam M, Sengara IW, Asrurifak M (2013) Proposed Sismic hazard map for conventional bridge code in Indonesia

Altamimi Z, Rebischung P, Métivier L, Collilieux X (2016) ITRF2014: a new release of the international terrestrial reference frame modeling nonlinear station motions. J Geophys Res: Solid Earth 121:6109–6131

Asrurifak, M., Irsyam, M., Aldiamar, F.: Seismic Hazard dan Pembuatan Seismic Hazard Map untuk Jembatan. (2012).

BNPB (2019) Data Informasi Bencana Indonesia. BNPB

Daryono MR, Natawidjaja DH, Sapiie B, Cummins P (2019) Earthquake geology of the lembang fault, West Java Indonesia. Tectonophysics 751:180–191

Federal Emergency Management Agency (FEMA) (2013) HAZUS-MH MR4 Technical Manual. National Institute of Building Sciences and Federal Emergency Management Agency (NIBS and FEMA), p 712

Irsyam M, Natawijaya DH, Daryono MR, Widiyantoro S, Asrurifak M, Meilano I, Triyoso W, Hidayati S, Rudiyanto A, Sabaruddin A (2017) Development of new seismic hazard maps of Indonesia 2017

Kokusho T, Fujita K (2002) Site investigations for involvement of water films in lateral flow in liquefied ground. J Geotech Geoenviron Eng 128:917–925

Natawidjaja DH (2018) Major bifurcations, slip rates, and a creeping segment of Sumatran Fault Zone in Tarutung-Sarulla-Sipirok-Padangsidempuan, Central Sumatra Indonesia. Indonesian J Geosci 5:137–160

Natawidjaja DH, Bradley K, Daryono MR, Aribowo S, Herrin J (2017) Late quaternary eruption of the Ranau Caldera and new geological slip rates of the Sumatran Fault Zone in Southern Sumatra Indonesia. Geosci Lett 4:21

PusGen (2017) Peta Sumber dan Bahaya Gempa Indonesia Tahun 2017. Puslitbang Perumahan dan Pemukiman, Bandung

Part IV
Adaptation to Climate Change-Induced Hazards

Chapter 13
Climate Change-Induced Geotechnical Hazards in Asia: Impacts, Assessments, and Responses

Kazuya Yasuhara and Dennes T. Bergado

13.1 Introduction

Climate change, in Intergovernmental Panel for Climate Change (IPCC) usage, is a change in the climate state that is identifiable (e.g. using statistical tests) by changes in the mean and/or variation of climate properties, and which persists for an extended period: typically decades or longer. Included is any change in climate over time, whether caused by natural factors or by human activity. By contrast, the United Nations Frameworks Convention on Climate Change (UNFCCC) regards climate change as solely attributed to human activities: climate variation refers to changes attributable to natural events.

Based on a review of the information presented above for Asia-Pacific regions (CRED 2016) this report firstly presents an overview of the present situation and future trends of geo-disasters in the context of climate change and presents possible adaptive measures against disasters. Subsequently emphasized is the importance of combined effects of plural events, which increase the probability of extreme events, triggering devastating consequences. The necessity of adaptive measures is highlighted for overcoming climate change-associated compound geo-disasters.

K. Yasuhara (✉)
GLEC, Ibaraki University, Ibaraki, Japan
e-mail: kazuya.yasuhara.0927@vc.ibaraki.ac.jp

D. T. Bergado
Asian Institute of Technology, Bangkok, Thailand

© The Author(s), under exclusive license to Springer Nature Singapore Pte Ltd. 2023
H. Hazarika et al. (eds.), *Sustainable Geo-Technologies for Climate Change Adaptation*, Springer Transactions in Civil and Environmental Engineering, https://doi.org/10.1007/978-981-19-4074-3_13

13.2 Recent Natural Disasters Caused by Climate Change

13.2.1 Brief Review of Asian Natural Disasters

After reviewing case histories of natural disasters in Asian regions, Kokusho (2005) described the following: (i) we might be overlooking extreme events that occurred in sparsely populated areas, especially in rural areas; (ii) earthquakes, tsunami events, floods, slides, volcanoes, surges, and wind storms will continue as salient causes of future catastrophes; (iii) increasingly expanding areas of human activities are creating conditions for compound hazards of new types leading to future extreme events. Kokusho also pointed out that new compound hazards are expected to result from (i) Asian population growth and economic development, (ii) new urban facilities produced with little experience of severe disasters, and (iii) other compound hazards such as land subsidence related flooding from increasingly vulnerable dykes.

From Kokusho's assertions, questions arose such as "Have those extreme natural events been caused by climate change including global warming?" The answer to this question has been clear since 2005: a decade and a half ago. In fact, IPCC reports (IPCC 2012, 2014) present evidence that climate change might trigger extreme disasters. Nevertheless, uncertainty related to this issue has persisted. Therefore, the following working hypothesis has been used to date: some events might result from climate change; others might not. Therefore, efforts at research and information exchange through international cooperation of professionals and stakeholders from related fields must be sustained, and particularly in Asia-Pacific regions.

Unfortunately, however, after 2005, no successive efforts have been conducted following the procedure undertaken by Kokusho. Therefore, the results from the statistical investigation of disastrous events in the last 27 years (CRED 2016) are briefly outlined herein for additional consideration of what we will be endangered by natural events in the coming decades, particularly in the Asia-Pacific region.

Figure 13.1 shows the regional percentage of geo-disasters including earthquakes, floods, slides, volcanos, waves/surges, and wind storms from 1990 through 2016 in terms of numbers of events, human fatalities, affected population, and economic

Fig. 13.1 Regional percentages of geo-disasters

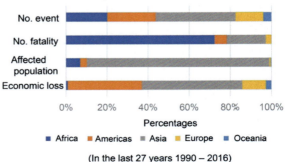

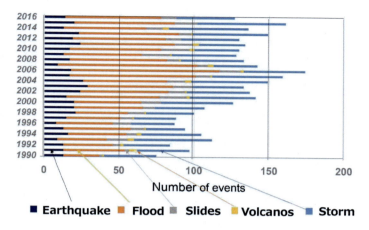

Fig. 13.2 Annual major events during the last 27 years in Asia (Kukusho, 2007)

loss. Similarly to earlier investigations by Kokusho, Asia currently overwhelms other regions in all aspects related to damage. That dominance is expected to persist.

Figure 13.2 presents variations of the number of disaster events in the last 27 years in Asia, indicating the following.

(i) The number of events during 1990–2006 is almost twice those occurring during 2007–2016.
(ii) The numbers of floods and storms occurring are closely related. Therefore, special attention must be devoted to their combination from the viewpoint of climate change.

In terms of human fatalities, earthquakes, volcanoes, and floods are major calamities, whereas landslides cause fewer fatalities, as depicted in Fig. 13.3. Earthquakes cause many casualties, probably because they sometimes produce tsunami waves, as occurred after the Sumatra earthquake in 2007 and in Tohoku-Off Pacific Earthquake in 2011.

Fig. 13.3 Percentages by deaths in Asia

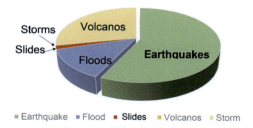

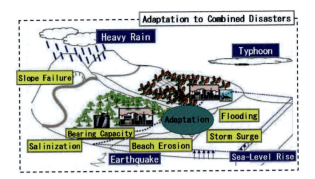

Fig. 13.4 Factors affecting compound disasters

13.2.2 Compound Disaster Importance (Yasuhara 2016a; Yasuhara et al. 2017)

Special attention must be devoted to compound disaster effects because these magnify disaster loss and damage. The following are necessary: (i) clarifying climate change induced natural disaster mechanisms; (ii) predicting future events and outcomes; (iii) estimating likely economic losses and costs related to precautions and preparedness; and (iv) proposing adaptation techniques and strategies.

Referring to Fig. 13.4, one can consider the nature of compound disasters. The following scenarios should be included in that consideration: (i) a second natural disaster occurs immediately before or after a major disaster, generating catastrophic consequences; (ii) damage is compounded through combinations of the natural disasters with a vulnerable background or human and social situations (see Fig. 13.5); and (iii) the psychological aftermath multiplies the damage. This classification accords well with that proposed by Kokusho (2005).

13.3 Recent Trends of Factors Triggering Sediment Disasters

13.3.1 Influence of Climate Change

Figure 13.6 shows that one can assume that climate change as defined in the *Introduction* produces (i) sea-level rise (SLR), (ii) magnification of and/or increase in the number of typhoons, (iii) variation of precipitation characteristics leading to torrential rainfall or extreme drought, and (iv) thermal variation of ground surfaces. This report specifically addresses items (i)–(iii).

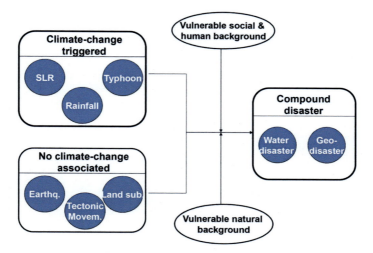

Fig. 13.5 Processes of compound disasters

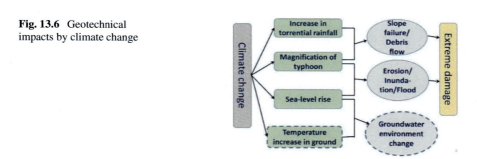

Fig. 13.6 Geotechnical impacts by climate change

13.3.2 Sediment Disaster and Precipitation (Yasuhara 2016a, Yasuhara et al. 2017)

Figure 13.7 depicts variations of sediment disaster occurred during the 15 years of 1999–2014 in Japan. Sediment disasters include slope failure, landslides, and debris flow. Figure 13.8 shows the annual frequency of torrential rainfalls of more than 50 mm/year for the last 35 years. Both figures show gradually increasing tendencies of both intensity and frequency in recent years.

When results of sediment disasters and torrential rainfall are compared, one obtains Fig. 13.9, which shows that the frequency of sediment disasters tends to increase nonlinearly over time.

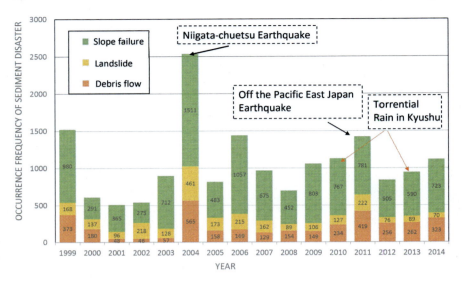

Fig. 13.7 Sediment disaster variation

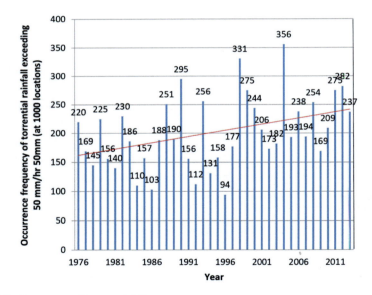

Fig. 13.8 Occurrence of heavy rainfall

13.3.3 Earthquake Tendency

Earthquakes represent another important factor triggering sediment disasters and geo-disasters. What would happen to human life if a great earthquake were to strike before

13 Climate Change-Induced Geotechnical Hazards in Asia … 203

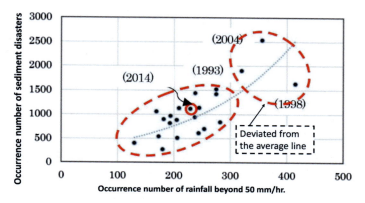

Fig. 13.9 Sediment disasters and heavy rainfall

or after a strong rainfall? This worst-case scenario would be a climate change associated compound geo-disaster. We propose preparedness by answering the question posed above, but ascertaining the correct answer is difficult.

To consider whether such a situation as that worst case scenario has actually occurred before, one can investigate the tendencies of large earthquakes over the last few decades. Figure 13.10 portrays variations in the occurrence frequency of earthquakes having lower 6 intensity over time from 1960 to the present. All data were obtained from the Japan Meteorological Agency. Figure 13.10 clarifies that the frequency of earthquakes with lower 6 intensity has been increasing over time, particularly since 1995. For that reason, one should take precautions for large earthquakes and heavy rainfalls. Preparation must be made for a worst case in which a climate change-induced heavy rainfall and a great earthquake might occur simultaneously or nearly simultaneously, although this worst case might be extremely rare.

In the same manner, as that used for Fig. 13.9, the influences of earthquakes with an intensity of lower than 5 or less on the frequency of sediment disasters are

Fig. 13.10 Occurrence of large earthquakes

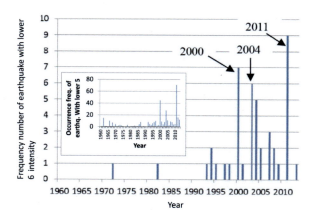

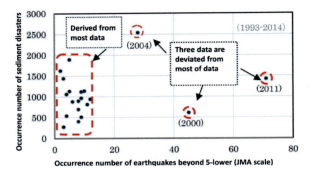

Fig. 13.11 Sediment disasters caused by tremors

presented in Fig. 13.11. Comparison with the result in Fig. 13.9 shows no mutual relation. Comparison of Figs. 13.9 and 13.11 suggest that torrential rainfall influences sediment disaster frequency more strongly than large earthquakes do.

13.3.4 Sediment Disaster Prediction

Studies conducted by Kawagoe et al. (2014) have assessed the prediction of future climate change induced sediment disasters and the economic losses that are likely to be sustained from those disasters.

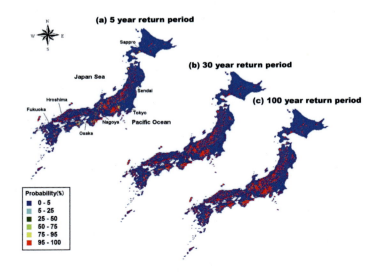

Fig. 13.12 Disaster probability increase (Kawagoe et al. 2014)

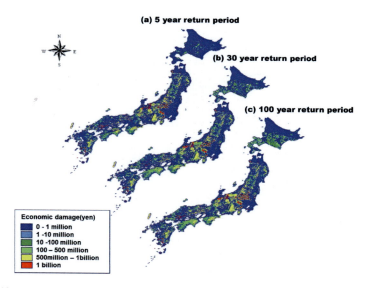

Fig. 13.13 Economic loss increase caused by disasters (Kawagoe et al. 2014)

Figure 13.12 exhibits the predicted results of sediment disasters, indicating that the probability of sediment disaster occurrence increases concomitantly with increasing duration of the return period of rainfall. This tendency is more apparent in western Japan than in eastern Japan. Nevertheless, the predicted economic losses presented in Fig. 13.13 are much greater in western areas than in eastern areas. The cumulative economic loss amount by the end of the twenty-first century is expected to be almost twice that of the present day. The results presented in Figs. 13.12 and 13.13 are mutually consistent.

13.4 Compound Disasters

13.4.1 Compound Disasters Related to Sea-Level Rise

13.4.1.1 Inundation Caused by Sea-Level Rise

Among the many coastal deltas around the planet, the Nile, Ganges, and Mekong are designated as extremely vulnerable coastal deltas (IPCC 2007). Maruyama and Mimura (2010) undertook numerical prediction of SLR effects on inundation that can be expected to prevail by the end of the 21t century. Results obtained with the expectation of no adaptation are depicted in Fig. 13.14, indicating that wide areas in Asian regions are expected to be inundated by SLR.

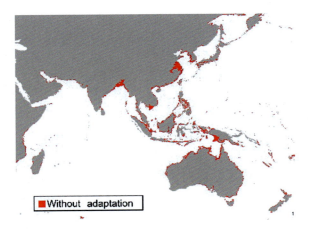

Fig. 13.14 Inundation predictions for 2100 for the SRES A1B scenario (dark shows inundation areas) (Maruyama and Mimura 2010; Mimura 2013)

13.4.1.2 Inundation Caused by SLR Combined with Land Subsidence

Relative vulnerability presented earlier in Fig. 13.14 shows no combined effect of SLR with land subsidence. To resolve this gap, an attempt was made to plot the representative locations experiencing severe land subsidence in Fig. 13.15 (Maruyama and Mimura 2010), which presents inundated areas obtained by assuming a sea-level rise to 88 cm at the end of the 21st century for the SRES A1B scenario (IPCC 2007).

Figure 13.15 (Mimura 2013) shows inundation areas in Southeast Asia, as influenced by land subsidence and SLR. This combined effect of land subsidence and SLR is expected to increase relative SLR, as presented in Fig. 13.15, which in turn

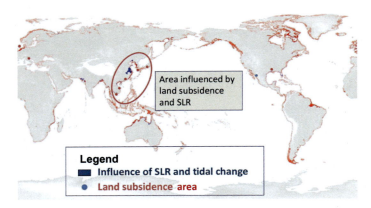

Fig. 13.15 Inundation areas with combined SLR and tidal change with land subsidence. Revised from Maruyama and Mimura (2010)

engenders increased inundation. Therefore, precise predictions of time-dependent variations of SLR and land subsidence should be conducted at least to 2100.

13.4.1.3 Case Study: Chao Phraya Delta, Thailand

(1) *Outline of the case study*

Deltas such as the Chao Phraya Delta, Red River Delta, and Mekong River Delta portrayed in Fig. 13.16 are known as "Mega deltas." An IPCC report (2007) described that Mega deltas are expected to be most vulnerable to natural disasters induced by global climate change because they are invariably located in very low land in coastal regions, which will be affected intensely and directly by sea-level rise caused by global warming. In actuality, land subsidence has occurred already in many Mega delta areas. A typical region that has been affected by both SLR and land subsidence is the Chao Phraya Delta in Thailand, which this study specifically examines.

To assess the influence of the dual impacts of sea-level rise and land subsidence on inundated areas of the Chao Phraya Delta, the future situation of land subsidence in 2100 has been predicted using a method of reliable land subsidence mapping based on observations of settlement proposed by Murakami et al. (2005). A future elevation model in the objective regions was produced by incorporating the present elevation model and the predicted land subsidence into GIS. To investigate the influences of dual impacts of both SLR and land subsidence on the inundation area in the Chao Phraya Delta, a hazard map of the inundation area has been produced. The hazard map shows the approximately 1,000 km² inundation area created by the dual impacts of SLR and land subsidence, which are explained later in greater detail.

(2) *Present situation of land subsidence in Chao Phraya*

During 1996–2003 in the Chao Phraya Delta settlement was monitored at approximately 748 observation locations. The present situation of land subsidence in the

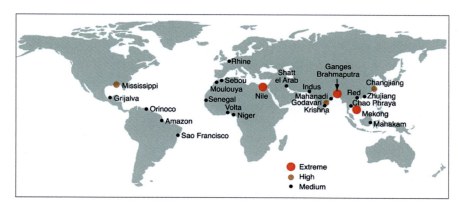

Fig. 13.16 Relative vulnerability of coastal deltas (IPCC 2007)

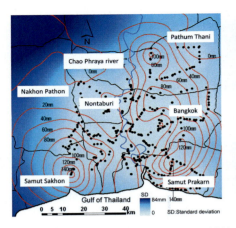 

(a) Actual land subsidence during 1996–2003 (b) Expected land subsidence in 2100

Fig. 13.17 Contour lines of land subsidence in the Chao Phraya Delta (Murakami and Yasuhara 2011)

objective regions can be elucidated using time-series records of settlement. In fact, a land subsidence map has been produced using reliable land subsidence mapping with a spatial interpolation procedure based on geostatistics proposed by Murakami et al. (2005), and Murakami and Yasuhara (2011) in this study. The mapping method can show not only the distribution of the expected settlement but also the distribution of estimated standard deviations based on the spatial correlation relations of settlement. The interpolation method is based on Kriging, a spatial interpolation method. Estimation at a location is assumed to be expressible as a linearly weighted summation of the observations. Figure 13.17a presents interpolated results of the distribution of land subsidence in the objective region during 1996–2003. The estimated settlement is drawn as contour lines. The estimated standard deviations are shown as raster data. The map shows that severe land subsidence has taken place in Samut Prakarn, the middle of Samut Sakhon, and the north of Pathum Thani (Murakami and Yasuhara 2011).

(3) *Assessing effects of dual impacts of SLR and land subsidence on inundation areas*

To assess the influence of dual impacts of sea-level rise attributable to global warming and land subsidence on inundation areas in the Chao Phraya Delta, GIS-aided spatial analysis was conducted based on the predicted results of future land subsidence in 2100, as presented in Fig. 13.17b (Murakami et al. 2005).

The inundation area, defined as a region with a ground level that is under sea-level in a future situation, was calculated from the present ground level considering future land subsidence. The SLR in this study was assumed as 59 cm according to an IPCC

13 Climate Change-Induced Geotechnical Hazards in Asia …

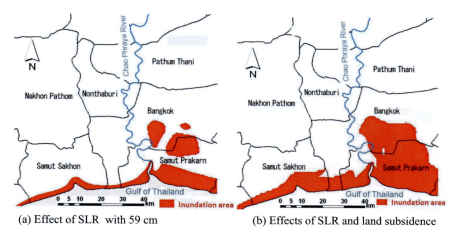

(a) Effect of SLR with 59 cm   (b) Effects of SLR and land subsidence

Fig. 13.18 Predicted inundation areas in the Chao Phraya Delta. From Murakami and Yasuhara (2011)

report (2007). Figures 13.18a, b present calculated results of inundation caused by SLR in 2100.

Figure 13.18a presents inundation regions considering SLR only. When considering the impact of SLR alone, results show that the coastal regions in Samut Sakhon, Bangkok, and Samut Prakarn and the middle region of Bangkok are inundated. Finally, Fig. 13.18b depicts inundated regions created by dual impacts of sea-level rise and land subsidence considering estimation error attributable to spatial interpolation and land subsidence prediction. The map in Fig. 13.18b shows that the inundated region expands markedly: nearly twice (1269 km^2) as much as in the case considering SLR alone (634 km^2).

Overall, results underscore the importance of precisely estimating the inundated region in a future situation affected by climate change to consider effects not only of sea-level rise but also of land subsidence, particularly through the devotion of careful attention to the estimation error.

13.4.2 Compound Disasters Related to Earthquakes (Yasuhara 2016a; Yasuhara et al. 2017)

An example pertaining to this compound disaster was experienced during the 2004 Niigata-Chuetsu Earthquake, which occurred after sustained rainfall lasting for nearly one month.

As shown in Photo 13.1, this disaster caused the collapse of highways and derailment of the *Shinkansen*, its first such accident ever, leading to slope failure at around

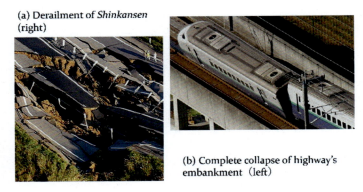

Photo 13.1 Damage to a highway and railway in 2004 Niigata Chuetsu Earthquake

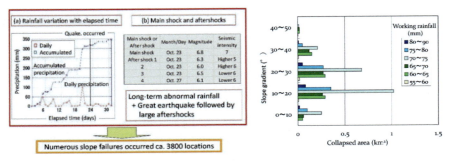

(a) Slope failures caused by combined effects (b) Precipitation characteristics [13]

Fig. 13.19 Typical compound disaster in 2004 Niigata Chuetsu earthquake

4000 locations. It was induced mainly by (i) saturation of unsaturated soils by sustained rainfall and (ii) frequent and powerful aftershocks (Fig. 13.19a).

One characteristic feature is that the amount of collapsed sediments becomes large when rainfall exceeds 70 mm Nunokawa et al. (2007) (Fig. 13.19b). Furthermore, the gentler the slope gradient becomes, the more marked its tendency becomes. Another characteristic mode of damage is that heavy snowmelt occurring after an earthquake induced secondary sediment disasters. Therefore, although the year had a historically large number of sediment disasters, as presented earlier in Fig. 13.11, this disaster can be regarded as a compound disaster. The combined effects of the earthquake with rainfall and snow melting before and after the earthquake created and exacerbated the disaster.

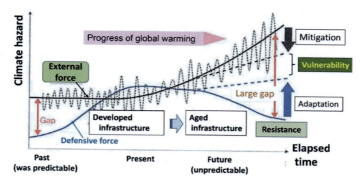

Fig. 13.20 Definition of resilience. Modified from Komatsu et al. (2013)

13.5 Response Measures: Geotechnical Adaptation Development

13.5.1 Resilience Against Climate Change-Induced Disaster Risks

In this paper, resilience is defined as the human potential able to fill in the gap separating the external force caused by climate change and the defensive force against climatic events, as presented in Fig. 13.20 (Komatsu et al. 2013). This gap is sometimes designated as "vulnerability": resilience mitigates the vulnerability.

13.5.2 From Rigid to Flexible Structures

13.5.2.1 Resilient Coastal Structures

One idea for weakening external forces such as wave actions is the replacement of concrete walls with geosynthetic-reinforced structures (GRS) with the addition of cement into soils wrapped by GS, or sewing GS into soil layers as presented in Fig. 13.21. This method can be assigned to a change of concept from rigid to flexible structures against intense external forces.

Model tests using the wave flume apparatus presented in Fig. 13.22 indicate that cement addition and sewing GS wrapping soil layers are effective for maintaining the stability of coastal wall structures assailed by wave action, as presented in Fig. 13.23.

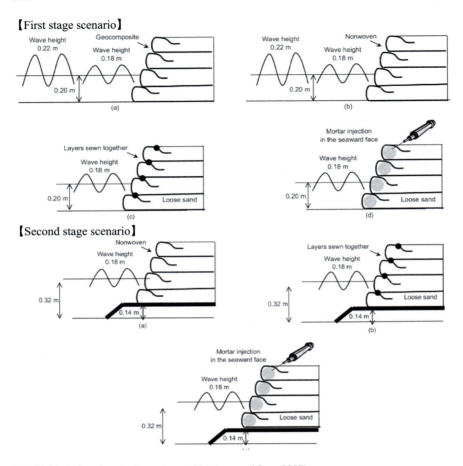

Fig. 13.21 Laboratory testing scheme (Yasuhara and Juan 2007)

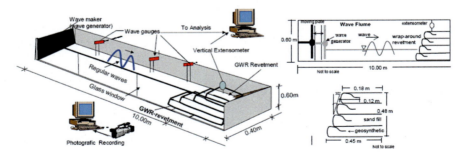

Fig. 13.22 Model test setup and GWR dimensions

13 Climate Change-Induced Geotechnical Hazards in Asia …

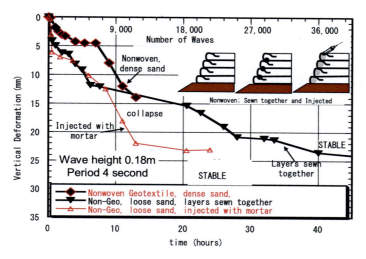

Fig. 13.23 Typical results of model tests (Yasuhara and Juan 2007)

13.5.2.2 Resilient Riverine Structures

Erosion and subsequent slope failure of mountain road embankments occur caused by the flooding of the adjacent river during the extreme rainy season. In particular, riverbank erosion tends to occur at the outside curve of the meandering river in Thailand. As a flexible adaptive measure against the erosion of this kind caused by severe rainfall, the application of geosynthetics as shown in Fig. 13.24 is effective for slope reinforcement.

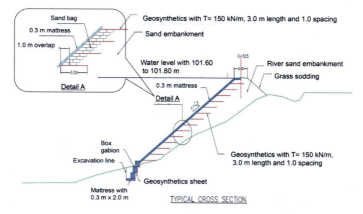

Fig. 13.24 Erosion control using geosynthetics. After Tanchaisawat et al. (2009)

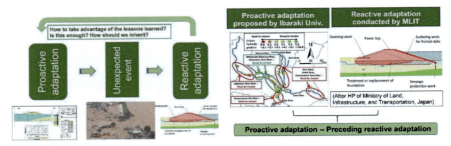

(a) Concept of proactive measure (b) Reactive and proactive measure to flood

Fig. 13.25 Example of transition from proactive to reactive measures against flooding

13.5.3 From Reactive to Proactive Measures

One important strategy is a philosophy of "From reactive to proactive measures" (Yasuhara et al. 2016b).

Figure 13.25 presents a key sketch for the concept on this philosophy from the experience of the flood damage sustained during the 2015 Kanto-Tohoku heavy rainfall having led to severe damage to infrastructure and fatalities (Yasuhara et al. 2016b).

As portrayed in Fig. 13.25, one achievement by Fujita et al. (2013) indicated which parts of main rivers in the Kanto regions are vulnerable to dyke erosion and which might therefore benefit from proactive adaptation efforts to river dyke erosion. Unfortunately, however, the results portrayed in Fig. 13.25b were not noticed before a river dyke damage event in 2015 simply because the investigation of the erosional characteristic of dyke soils was conducted before the devastating flood event of 2015 in the Kinu River (Komatsu et al. 2013).

The concept of reactive to proactive measures is equivalent to the concept of "adaptive adaptation" proposed by Yasuhara and Juan (2007).

13.5.4 Monitoring Importance

For slopes, which differ slightly from river dykes, adopting a monitoring system is fundamentally important for combining precipitation characteristics and information from slope surface movements, water contents, and pore pressures in soil deposits that form the slopes (Fig. 13.26) (Yasuhara and Juan 2007).

A powerful proactive measure is the utilization of information communication technology (ICT). It can be particularly effective for monitoring infrastructure behavior before a climatic event.

In line with this concept, large-scale model tests for slope failure caused by rainfall were conducted at the National Research Institute for Earth Science and Disaster

Fig. 13.26 Slope stability monitoring system

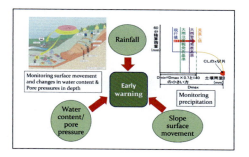

Prevention (NIED) (Dairaku et al. 2014). Photo 13.3 shows a slope outline with a 30° gradient, which consists of Masa soil with an initial water content of around 7.9%. Masa soils used for the model tests closely resemble soils involved in the debris flow that occurred in Hiroshima in August 2014 (JSCE and JGS 2014).

To ascertain how the variation of the acceleration on sloping ground with the intensity of rainfall is correlated with slope failure, sensing IC-tags for acceleration measurement, called MEMS, were installed as shown in Photo 13.2 together with equipment to measure volume moisture content variation.

The results depicted in Fig. 13.27a present that natural frequency measured from IC-tags tends to decrease concomitantly with increasing unit volume weight of Masa soils consisting of slopes. In other words, the natural frequency in soils decreases concomitantly with increasing water contents caused by successive rainfall.

Actually, different thresholds of natural frequency or critical volume moisture content should be ascribed to differences in sensitivity to rainfall and water retention depending on the soil type, as presented in Fig. 13.27b as a key sketch (Yasuhara 2016a).

Photo 13.2 Large-scale model tests for slope undergoing rainfall

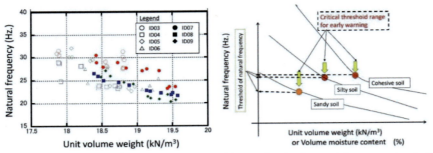

(a) Natural frequency to unit volume weight (b) Key sketch of natural frequency threshold

Fig. 13.27 Acceleration characteristics of soil slope undergoing rainfall (Dairaku et al. 2014)

13.5.5 Synergy of Mitigation with Adaptation

As the most expected mitigation measure to reduce GHG emissions, IPCC has proposed "Geo-engineering," which includes solar radiation management (SRM) and carbon dioxide removal (CDR) (IPCC 2007).

Although less mitigation measures for geo-disasters were described, an example of potential mitigation measures is introduced here. Fundamental research conducted by Umino et al. (2017) presents carbon dioxide (CO_2) fixation properties of iron and steel slag containing calcium. To investigate those CO_2 fixation properties, CO_2 fixation tests were conducted with constant flow. Results show that when the CO_2 concentration 4500 L L$_{-CO2}$/L flowed in a specimen by 0.05 L/min, for a non-aged steelmaking slag, the amount of fixed CO_2 was the maximum: 0.04 g$_{-CO2}$/g$_{-slag}$. The amount of CO_2 fixed in the steelmaking slag resulted from about 20% of soluble calcium in the chemical reaction. The CO_2 fixation mechanism is presented in Fig. 13.28.

Therefore, the quantity of CO_2 fixation can be evaluated in terms of its CO_2 fixation mechanism using the quantity of water-soluble calcium. Figure 13.28 shows that results should be obtained in a practical manner to contribute to geotechnical applications for the formation of a sound material-cycle society and a low-carbon society.

13.5.6 Combined Green-Infrastructure with Grey-Infrastructure

13.5.6.1 Vietnamese Case

Vegetation using grasses and trees is sometimes useful for protecting infrastructure from climate change together with maintaining landscapes at coasts and riverine areas, sometimes in the case of combining the concrete structures. Cases are presented

13 Climate Change-Induced Geotechnical Hazards in Asia …   217

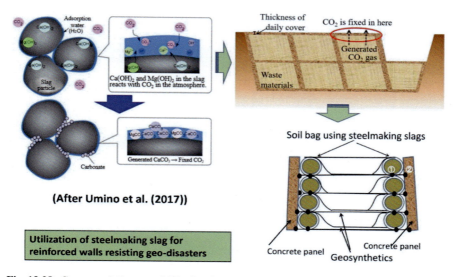

Fig. 13.28 Conceptual diagram of CO_2 fixation mechanism. From Umino et al. (2017)

in Fig. 13.29 where Malaleuca is combined with a sand mattress and Fig. 13.30 in which mangroves are combined with concrete and palm fibers are mixed with soil and cement for dykes.

The techniques consequently produce multiple protection facilities that incorporate the placement of natural and artificial measures and improvement of the dykes themselves.

Fig. 13.29 Combination of Malaleuca with sand mattress

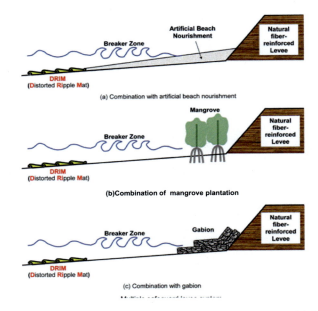

Fig. 13.30 Multiple protection placement of natural and artificial measures and dyke improvement (Yasuhara 2016a)

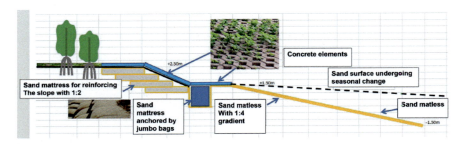

Fig. 13.31 Mattress foundation with vegetation for coastal protection. From Van (2018)

The green technique for river dykes is now being extended to protective measures for coastal dykes at Tỉnh Tiền Giang Province in the Mekong Delta area of Vietnam, as depicted in Fig. 13.31.

13.5.6.2 Bangladesh Case Study of Jute Usage

Regarding the case of applying locally available materials, Matsushima et al. (2010) presented a case study of the use of jute inclusions in soils used for local agricultural road embankments to resist erosion possibly caused by climate change, as depicted in

Fig. 13.32 Successful case study for road embankment using jute (Matsushima et al. 2010)

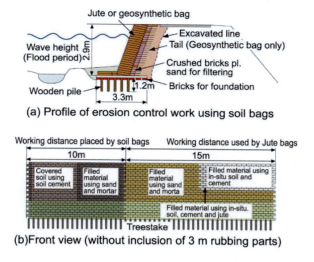

Fig. 13.31. This case occurred on local roads for agricultural purposes in Bangladesh. It represents a good practice in Southeast Asian countries against climate change-triggered events (Fig. 13.32).

13.5.6.3 Case Study of Combined Usage of Bamboo with Concrete Walls

Bamboo is distributed widely not only in Japan but also among many Southeast Asian countries. Unfortunately, less development of utilization is devoted to such civil engineering practices except the case in which the reinforced earth, as depicted in Fig. 13.33, is the only case in Japan for use of split bamboo sheets with 60 mm width as strips for reinforced materials of wall structures instead of conventionally used metals and geosynthetics.

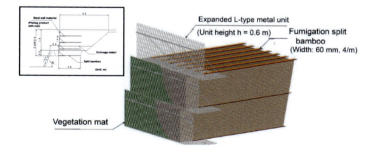

Fig. 13.33 Bamboo sheet used retaining walls (from Kyousei Co. Ltd.)

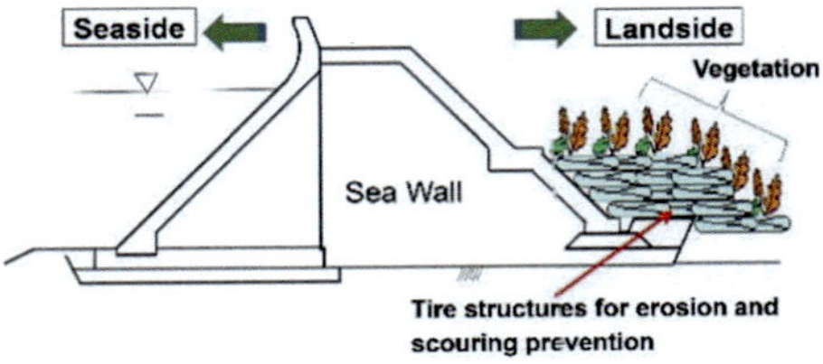

Fig. 13.34 Development of used tires for coastal protection together with vegetation (Hazarika and Fukumoto 2016)

Fig. 13.35 Shredded tires wall (Tanchaisawat et al. 2009)

13.5.7 Utilization of Industrial By-Products

As depicted earlier in Fig. 13.28, the use of steel slag capable of absorbing CO_2 is an example of the effective utilization of industrial by-products. Another example can be given in which disused tire structures with incorporated vegetation are adopted for protecting dykes undergoing severe storms followed by overlapping the dyke as depicted in Fig. 13.34. Unfortunately, this typical combined usage of green-infrastructure with grey-infrastructure has not yet been proved both in laboratories and in the field.

Along with the usage of whole disused tires, shredded used tires of different sizes, sometimes wrapped around by geosynthetics (GS), are used as backfill materials for reinforced wall structures, as depicted in Fig. 13.35.

13.6 Recommendation from Geotechnical Engineering

Although increased activities have highlighted the roles of geotechnical engineering in climatic events, as shown in Table 13.1, no knowledge and findings have been included in the IPCC AR5, which was published in 2014 (IPCC AR5 2014). Concretely speaking, one can find no mention of the glossary and index in AR5 of geotechnical terminologies such as landslides or land subsidence although the terminology of *landslides* appeared in the special report on disasters: SREX (IPCC 2012).

The situation stated above is ascribed to the following:

13 Climate Change-Induced Geotechnical Hazards in Asia ...

Table 13.1 Some activities of climate change associated geotechnical issues

From	Fear EToH	Activity	Country	Reference
2002	2012	Research as CoE on Climate Change- induced Geo-disasters	Norway	Carried out at Norwegian Geotechnical Institute (NGI). The representative is Dr. Farokh Nadhim
2005	2009	Strategic project on climate change adaptation called S-4	Japan	Supported by Ministry of Environment
2010	2014	Strategic project on climate change adaptation called S-8	Japan	Supported by Ministry of Environment
2013	Present	Asian Technical Committee (ATC1) on Geotechnical Mitigation and Adaptation to Climate Change- Induced Geo-disasters in Asia-Pacific Regions	Asia	Activity of ATC1 is included in International Society of Soil Mechanics 8i Foundation Engineering
2013	2014	Research Committee on Synergy of Geo-disaster Risk management and Climate Change Adaptation	Japan	Organized by Japan Geotechnical Society (JGS) and chaired by Professor Hemanta Hazarika (Kyushu University)
2014	2015	Course on Global Warming-induced Geo-enviroments and Geo-disasters	Japan	Journal of Japan Geotechnical Engineering (in Japanese)
2016	2016	56th Rankine Lecture entiled Geotechnlcs, Energy and Climate Change	UK	Delivered by Professor Richard Jardine (Imperial College) under Organisation by British Geotechnical Association
2016	2016	Symposium on Climate Change-induced Geo-disaster Risk in Snowy Regions	Japan	Organized by Hokkaido Branch of JGS which was chaired by Professor Tatsuya Ishikawa of Hokkaido University
2017	2017	2017 PGS Workshop &15th G. A. Leonards Lecture on Climate Change & Geotechnical Engineering	USA	Held at Purdue University

(continued)

(i) No researcher or engineer of geotechnical engineering is noted as an expert among Coordinating Lead Authors (CLA), Lead Authors (LA), Contributing Authors (CA), or Review Editors (RE), at least in AR6.

(ii) Existence of geotechnical engineering has not been recognized in the academic field, particularly in the IPCC domain. Similarly, no geotechnical disaster or geo-disaster has been recognized as terminology.

The author proposes the following to improve this situation.

Table 13.1 (continued)

From	Fear EToH	Activity	Country	Reference
2018	2018	Special Issue on Climate Change & Geotechnical Engineering	Thailand	Journal of Southeast Asian Geotechnical Society (SEAGS) & Association of Geotechncal Societies in South East Asia (AGSSEA), Vol.47, No. 1
2019	Present	Colloborative Geotechnical Research Group in Branches of Hokkaido & Kyushu	Japan	Under the support from Geotechnical Research Committee in Japanese Society of Civil Engineers (JSCE)
2020		1st International Conference on CONSTRUCTION RESOURCES FOR ENVIRONMENTALLY SUSTAINABLE TECHNOLOGIES	Japan/UK	Climate change adaptation is a main topic

(i) International Society of Soil Mechanics and Foundation Engineering (ICSMGE) formulated a strategy of increasing awareness of the potential roles of geotechnical engineering in disaster prevention and mitigation through activities to contribute to the IPCC.

(ii) Encourage submission of highly qualified papers to highly reputed journals related to climate change, such as *Climatic Change, Mitigation and Adaptation Strategies.*

(iii) Introduce case studies from such regions as Kyushu and Hokkaido in Japan, and such countries as those in Southeast Asia, which are vulnerable to climate change

13.7 Conclusion

Climate change affects many aspects of human life on Earth. Although the controversial discussion has persisted on whether climate change is triggered by global warming, little room remains for doubt that extreme climatic events have been increasing worldwide. Particularly, the Asia-Pacific region is regarded as being among areas that are particularly vulnerable to climate change. In this sense, geotechnical engineering is expected to contribute to finding solutions to difficult issues confronting human beings. Unfortunately, however, no international attempt has been made to collect achievements related to geotechnical disasters, which can contribute to the development of technologies and policies for disaster risk reduction.

This study was conducted to present knowledge and lessons learned from several case studies in different countries in Asia. Based on this knowledge and lessons, special emphasis have been placed on the importance of tackling compound disasters, which include events combined with and without climate change-associated and non-associated factors. For preparation and countermeasures to be pursued in the future, we must acknowledge the following techniques and policies.

(i)   Greater insight should be sought in relation to proactive rather than reactive measures which belong to the concept called *"adaptive adaptation"*.
(ii) Synergetic approaches should be emphasized to combine adaptation with mitigation.
(iii) Green infrastructure should be explored to combine such grey infrastructures such as concrete structures and reinforced structures.

As stated above, this review can be anticipated as a milestone marking greater efforts by geotechnical engineers to contribute to international organizations studying climate change, such as the IPCC and COP.

Acknowledgements The author expresses his sincere gratitude for the financial support received as a grant-in-aid for the MEXT project entitled "Southeast Asia Based Climate Change Adaptation Network," of which a representative is Professor Tetsuji Itoh, Director of ICAS, Ibaraki University. In addition, cooperation from members of ATC1 in ICSMGE on "Geotechnical Mitigation and Adaptation to Climate Change-induced Geo-disasters in Asia-Pacific Regions (GMACC)" is highly appreciated. Statistical analysis of Asian disasters from CRED is credited to the great effort and contribution by Ms. Saori Fukushi, formerly Graduate Student at the Graduate School of Science and Engineering, Ibaraki University, Japan. Dr. Tomohiro Ishizawa, Chief Researcher of the National Research Institute for Earth Science and Disaster Resilience (NIED), kindly provided us analysis results from CRED data (1989–2005) to contribute to the SOA report given by Dr. Takeji Kokusho, at that time Professor of Chuo University, Japan, and currently Professor Emeritus at that university.

References

CRED, Annual disaster statistical review (CRED ADSR) 2006–2015. http://www.cred.be/
Dairaku A, Murakami S, Komine H, Saito O, Ishizaki, Maruyama I (2014) Vibration measurement using MEMS IC sensing tags in slopes during and after rainfall. In: Proceedings of the 50th conference on JGS, CD-ROM (in Japanese)
Fujita K, Komine H, Murakami S, Yasuhara K, Taniguchi Y (2013) Widely ranged river dyke erosion estimation. In: Ninth symposium on environmental geotechnical engineering, pp 217–222 (in Japanese)
Hazarika H, Fukumoto Y (2016) Sustainable solution for Seawall protection against Tsunami-induced damage. Int J Geomech ASCE. https://doi.org/10.1061/(ASCE)GM.1943-5622.0000687
IPCC (2007) AR4. https://www.ipcc.ch/assessment-report/ar4/
IPCC (2012) Special report on managing the risks of extreme events and disasters to advance climate change adaptation: SREX. Cambridge University Press
IPCC (2014) AR5. Cambridge University Press. http://www.ipcc.ch/report/ar5/
JSCE and JGS (2014) Urgent Report for 2014 Hiroshima Sediment disasters caused by Torrential Rainfall, 2nd edn (in Japanese)
Kawagoe S, Esaka Y, Ito K, Hijioka Y (2014) Estimated sediment hazard damage using general circulation model outputs in the future. J Jpn Soc Civil Eng Ser. G (Environ Res) 70(5):I_167–I_175 (in Japanese)

Kokusho T (2005) Extreme events in geohazards in Asia. In: Proceedings of the international conference on geotechnical engineering for disaster mitigation & rehabilitation, Singapore, pp 1–20

Komatsu T, Shirai N, Tanaka M, Harasawa H, Tamura M, Yasuhara K (2013) Adaptation Philosophy and strategy against climate change-induced geo-disasters. In: Proceedings of the Tenth JGS symposium on environmental geotechnics, Tokyo, Japan, pp 76–82

Mimura N (2013) Sea-level rise caused by climate change and its implications for society—Review. Proc Jpn Acad Ser. B 89(7):281–301

Maruyama Y, Mimura N (2010) Global assessment of climate change impacts on coastal zones with combined effects of population and economic growth. Selected papers of environmental systems research, JSCE 38:255–263 (in Japanese)

Matsushima K, Mohri H, Nakazawa K, Yamada K, Hori T, Ariyoshi M (2010) A pilot study against wave-induced erosion at the rural roads in Bangladesh. J Geosynth 25:99–106 (in Japanese)

MurakamiS, Yasuhara K, Suzuki K (2005) Reliable land subsidence mapping by a geostatistical spatial interpolation procedure. In: Proceedings of the 16th international conference on soil mechanics and geotechnical engineering: geotechnology in harmony with the global environment, vol 4, pp 2829–2832

Murakami S, Yasuhara K (2011) Inundation due to global warming and land subsidence in Chao Phraya Delta. In: Proceedings of the 14th Asian regional conference SMGE, vol 1 (CD-ROM)

Nunokawa N, Murakami S, Yasuhara K, Komine H, Tsuchida A (2007) Effects of rainfall on slope failure caused by Niigata Chuetsu earthquake. In: Proceedings of the symposium on recent development in prediction and countermeasures for slope failure, Kyushu Branch of JGS, pp 221–224 (in Japanese)

Tanchaisawat T, Bergado DT, Voottipruex P (2009) 2D and 3D simulations of geogrid reinforced lightweight embankment on soft clay. Geosynth Int 16(6):420–432

Umino M, Komine H, Murakami S, Yasuhara K, Setoi K, Watanabe Y (2017) Iron and steel slag properties and mechanisms for carbon dioxide fixation in a low-carbon society. Geotech Eng J SEAGS & AGSSEA 48(1). ISSN 0046-5828

Yasuhara K, Juan R (2007) Geosynthetic-wrap around revetments for shore protection. Geotext Geomembr 10(1):1–12

Yasuhara K (2016a) Geotechnical responses to natural disasters and environmental impacts in the context of climate change. In: Phung (ed) Proceedings of the international conference on geotechnics for sustainable infrastructure development—Geotec Hanoi 2016a, pp 957–981

Yasuhara K, Murakami S, Koarai M, Ito T, Tsutsui K (2016b) Lessons learned from flood damage in Kinu River during the Kanto-Tohoku localized torrential downpour in 2015. In: Proceedings of the international symposium Hanoi geoengineering, Hanoi, Vietnam, pp 16–23

Yasuhara K, Kawagoe S, Araki K (2017) Geo-disasters in Japan in the context of climate change. Geotech Eng J SEAGS & AGSSEA 48(1):1–11. ISSN 0046-5828

Chapter 14
Effect of Vessel Waves on Riverbank Erosion: A Case Study of Mekong Riverbanks, Southern Vietnam

Tran The Le Dien, Huynh Trung Tin, Bui Trong Vinh, Trang Nguyen Dang Khoa, and Ta Duc Thinh

14.1 Introduction

An Giang province is located at the upstream of the Mekong river system flowing into the Vietnam territory where there are three large rivers flowing through, such as the Tien, Hau and Vam Nao rivers. The direction of Hau and Tien rivers are parallel from the Cambodia border to the sea area via Tran De and Dinh An estuaries. A segment of the Hau river that runs through An Giang province with about 100 km length, brings many benefits such as: the source of aquatic products, building materials, important transportation and flood relief routes.

The instability of Hau riversides has been occurring in Quaternary sediments complicatedly. Specially, the formation and development of landslide phenomenon that causes instability of Hau riversides and endangers human life, is technical and periodic with some specific mechanisms. Figure 14.1 shows the research area.

The riverbank erosion occurs at the surface layer due to impacts of natural waves resonating with vessel waves that create erosion pressure greater than the surface erosion resistance of the upper layer. Another reason for bank erosion is the physical feature of the ground layer. The surface layer will be bellied and shrunk because of level water differences by season and tide. That is one of the reasons causing cracks in ground and erosion of the riverbank under the impacts of high flow velocities (Le

---

T. T. Le Dien (✉) · H. T. Tin · B. T. Vinh · T. N. D. Khoa
Ho Chi Minh City University of Technology – VNU-HCMC, Ho Chi Minh City, Vietnam
e-mail: dien.tran@vnrentop.com

T. T. Le Dien
Rentop Corp, Ho Chi Minh City, Vietnam

T. D. Thinh
Hanoi University of Mining and Geology, Ha Noi, Vietnam

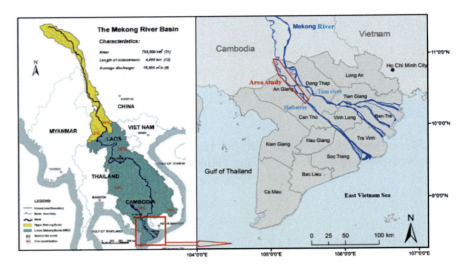

Fig. 14.1 Location of research area

Dien et al. 2012). In recent years, upstream of the isle is always eroded with a medium velocity of 10 m/year; especially some places have velocity up to 30 m/year (Fig. 14.2).

Hau River is an important arterial traffic through An Giang province that actuates social-economic development and increases vessel density on the river. The increasing vessel density and velocity created waves that impact banks consecutively and cause riverbank erosion seriously. Figure 14.3 indicates the number of vessels transporting through Binh Thanh islet and My Hoa Hung islet.

The geological structure of the Binh Thanh islet and My Hoa Hung islet is shown in Fig. 14.4.

The silty-clay on the top of the structure has a grain size distribution shown in Table 14.1. Besides, all the vessels wave essentially impact on this layer creating

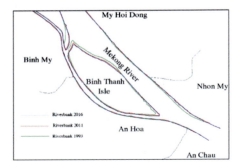
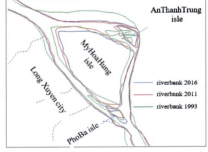

Fig. 14.2 Riverbank changes in recent year at My Hoa Hung and Binh Thanh islets, Hau river, southern Vietnam

14 Effect of Vessel Waves on Riverbank Erosion …

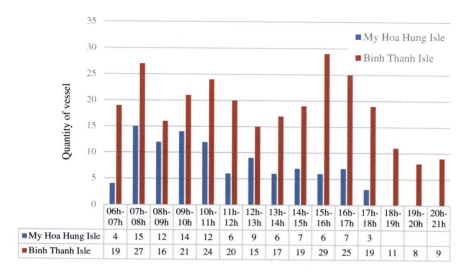

Fig. 14.3 The column chart shows the number of vessels transporting through Binh Thanh islet and My Hoa Hung islet

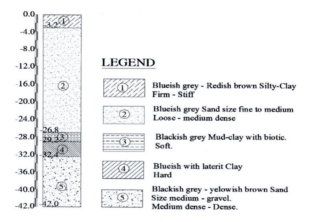

Fig. 14.4 Graphic log of My Hoa Hung and Binh Thanh islets, Hau river, southern Vietnam

Table 14.1 Analysis result of grain size and moisture content

No	Location	Sampling depth (m)	Grain size distribution (%)		
			Sand	Silt	Clay
1	My Hoa Hung isle	0.8–1.0	10.3	41.1	48.5
2	Binh Thanh isle	1.5–1.7	11.4	41.3	47.3

erosion, so, that is a reason why our research focuses on the silty Clay layer from the surface to −3.5. The value of grain size distribution is based on 33 samples collected at two Binh Thanh islets and My Hoa Hung islet.

In this paper, the authors have investigated the impact of vessel waves on Hau riverbank erosion and instability by vessel wave monitoring, erosion resistance experiment of riverbank materials and laboratory experiments.

14.2 Research Methods

14.2.1 Monitoring the Ship-Generated Waves

To evaluate the potential impacts of the vessel waves, a field observation has been carried out. The objectives of the observation are to (1) survey waterway traffic in selected areas; (2) review the bank erosion and (3) measure waves generated by vessels. Figure 14.5 sketches out the experimental model to monitor the vessel waves. The model consists of a wave height meter and an IR camera that allows recording the wave motion both day and night. A computer system with wireless Internet connection is also provided to remotely control the system.

The site monitoring for vessel waves has been using the AWH 10-USB at My Hoa Hung and Binh Thanh islets according to Table 14.2 and Fig. 14.6.

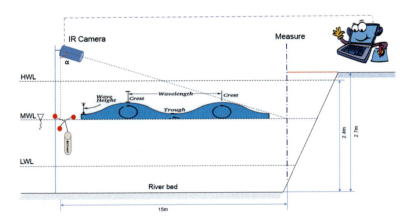

Fig. 14.5 Model of wave observation system

Table 14.2 Location for monitoring vessel waves

No	Locations	WGS 84		Monitoring period
		Lat	Long	
WMS1	My Hoa Hung isle	10°26′11.63″N	105°24′44.10″E	June 04th–11th, 2017
WMS2	Binh Thanh isle	10°28′45.43″N	105°20′28.65″E	Dec 5th–16th, 2017

Fig. 14.6 Monitoring vessel waves at My Hoa Hung and Binh Thanh islets

14.2.2 Experimental Jet Test

According to Hanson et al. (2004), the rate of erosion ε (cm/s) is proportional to the effective shear stress in excess of the critical shear stress as:

$$\varepsilon = k_d(\tau_w - \tau_c) \qquad (14.1)$$

where

k_d is the edibility coefficient (cm^3/N s)
τ_w is the applied shear stress (N/m^2)
τ_c is the critical shear stress (N/m^2)
τ_c, k_d is determined from the Jet test (Fig. 14.7).

$$\tau_c = \tau_0 * \left(\frac{J_p}{J_e}\right)^2 \qquad (14.2)$$

$$\tau_0 = C_f * \rho * U_0^2 \qquad (14.3)$$

    Based on grain size distribution in the analyzed result at My Hoa Hung and Binh Thanh islets, the remolded samples with sand-silt-clay content in each remolded sample were the same with 10% sand, 40% silt, and 50% clay. The moisture contents of these samples were changed from 44.0 to 55.1%. Remolded samples are shown in Fig. 14.8 and Table 14.3.

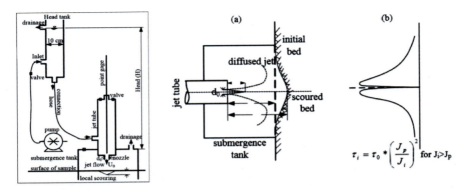

Fig. 14.7 Model jet test (Hanson and Cook 1997 and 2001)

Fig. 14.8 Remolded samples before testing (Vinh, 2009)

Table 14.3 Analysis result moisture content from remolded samples

No	Samples	Sand (%)	Silty (%)	Clay (%)	Moisture content (%)
1	SS10	10	40	50	44.0
2	SS11	10	40	50	45.1
3	SS12	10	40	50	47.8
4	SS13	10	40	50	50.6
5	SS14	10	40	50	52.8

The main causes of the phenomenon of erosion (τ_w) is natural waves and vessel waves pressure impact directly riverbank (Fig. 14.9).

According to Voulgaris et al. (1995), Althage J (2010) and Roulund et al. (2016):

$$\tau_w = \frac{1}{2}\rho f_w U_{\text{orb}}^2 \qquad (14.4)$$

where

τw Wave-average bed shear stress [N m^{-2}]
ρ Density of water [kg m^{-3}]
f_w Wave friction factor

14 Effect of Vessel Waves on Riverbank Erosion …

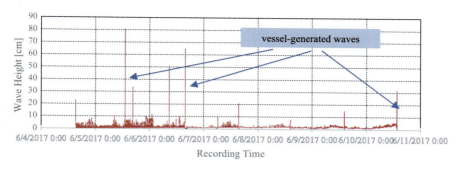

Fig. 14.9 Monitoring result of wave at My Hoa Hung isle

U_{orb}^2 Wave orbital velocity near the bed [m s^{-1}].

According to Soulsby et al. (1988)

$$U_{orb}^2 = \frac{\pi H}{T \cdot \sin(2\pi h/L)} \quad (14.5)$$

where

U_{orb}^2 Wave orbital velocity near the bed [m s^{-1}]
H Wave height [m]
T Wave period [s]
L Wave length [m^{-1}]
h SWL depth [m].

14.3 Analysis Results

14.3.1 Vessel Waves

Wind waves and vessel waves were observed at the monitoring sites. Figure 14.10 indicates the monitoring results of waves at My Hoa Hung isle (MWS1). The result shows that wind waves in this area were about 0.05–0.2 m in height. When vessels passed, wave heights increased significantly (0.3–0.6 m), especially during high tide and windy combination periods (0.8 m). High waves generated by vessels during high water level periods prevail at high frequency.

On the other hand, river segments running through Binh Thanh isle (left branch) are narrow, so impacts of waves and wind are so low (approx. 0.1 m). Monitoring results show that waves at MWS2 were mainly caused by vessels accounting for 77.04%. Two main reasons were investigated (1) Rach Goc ferry station, and (2) the main route of vessels to Long Xuyen city. Table 14.4 presents the statistics result of vessel wave frequency at monitoring locations.

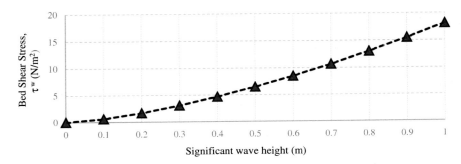

Fig. 14.10 Relationship between surface shear stress and wave height

Table 14.4 Distribution of frequency characterization of wave height

Wave height (mm)	MWS 1		MWS 2	
	Frequency	%	Frequency	%
0.01–10	2838	99.58	8382	22.96
10.01–20	5	0.18	17,276	47.32
20.01–30	2	0.07	6428	17.61
30.01–40	2	0.07	3075	8.42
40.01–50	0	0.00	1347	3.69
>50.01	3	0.11	0	0.00
Total	2850	100.00	36,508	100.00

Figure 14.10 shows the relationship between wave height and bed shear stress (BSS). The value of wave height is proportional to bed shear stress. When the BSS is higher than τ_c, the wave height at that BSS value will make the riverbank eroded.

14.3.2 Analysis Result of Jet Test

Analysis results in Jet Test for remolded samples with 10% sand, 40% silt and 50% clay are shown in Tables 14.5 and Fig. 14.11. When the moisture content increased from 44.0 to 45.1%, the τ_c decreased from 13.42 to 12.81 N/m². However, when the moisture content increased from 45.1 to 47.8%, the τ_c decreased suddenly from 13.42 to 2.80 N/m² and got a value of 0.02 at 55.1% moisture content. It can be said that the τ_c is inversely proportional to the moisture content, and at the liquid limit (47.8%) the τ_c will be decreased rapidly and approaches 0.0 N/m².

The results of the Jet test show that in the same grain size distribution, the increase of moisture content of soil reduces the critical shear stress, and when the moisture content is more than 55% the Critical shear stress is close to 0 N/m².

In Fig. 14.12, there were a total of 17 samples at Binh Thanh and 12 samples at

Table 14.5 Analysis results τ_c and k_d according to Jet test

No	Sample ID	Grain size distribution (%)			Moisture content (%)	τ_c (N/m^2)	k_d (m^3/N s)
		Sand	Silt	Clay			
1	SS10	10	40	50	44.0	13.416	0.126
2	SS11	10	40	50	45.1	12.812	0.168
3	SS12	10	40	50	47.8	2.803	0.523
4	SS13	10	40	50	50.6	0.280	0.536
5	SS14	10	40	50	52.8	0.185	1.810
6	SS15	10	40	50	55.1	0.019	2.510

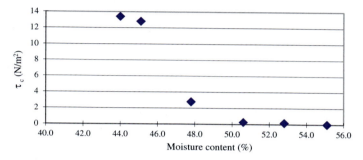

Fig. 14.11 Relationship between the τ_c and k_d moisture content of remolded samples

Fig. 14.12 Collect samples at My Hoa Hung and Binh Thanh islets

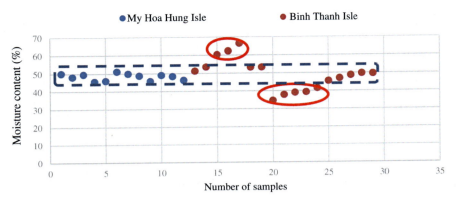

Fig. 14.13 Analysis result Moisture content of ground surface at My Hoa Hung and Binh Thanh isles

the MHH islet collected to calculate soil moisture content. The moisture content of samples from two islets was used to estimate the critical shear stress from the results of the Jet test.

Meanwhile, Fig. 14.14 shows that, the majority moisture content is from 45 to 55%. However, in the Binh Thanh islet, there are two groups of samples in two different moisture ranges including below 45% and upper 55%.

τ_c is a value shown the critical shear stress, thus, τ_c higher the sample easily destroyed and τ_c is an inverse ratio to the moisture content (Fig. 14.11). Therefore, the moisture content value of the sample collected on the field will be a parameter to compare with the shear stress of wave impact on the surface of riverbanks.

In Fig. 14.13, the majority of samples have moisture content in the range of 45–55% means the riverbanks are more easily destroyed if the shear stress of waves applying to. Besides, in Binh Thanh islet. There are three samples that have moisture content value higher than 55% means the τ_c value approximately equal to 0.

14.4 Discussion

Hau river is an important arterial traffic through An Giang province that actuates social-economic development and increases vessel density on Hau river. The increasing vessel density and velocity created wave trains consecutively that impact banks. These wave trains with a height under 1.0 m generated pressure impacting blue-grey and red-brown clay layers directly. This effect makes texture breakdown of river slope, erosion of materials and transportation of materials by the flow. This process takes place continuously that causes erosion of riverbanks seriously. The pressure of waves that impact riverbanks directly, is the shear stress of waves impacting riverbanks. Figure 14.14 shows the results of site experiments at My Hoa Hung and Binh Thanh isles. These experimental results show that blue-grey and red-brown

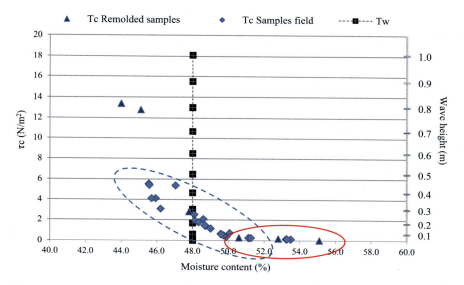

Fig. 14.14 Relationship chart between shear resistance of ground and shear stress of boat wave at My Hoa Hung and Binh Thanh islets, Hau river, southern Vietnam

clay layers with depths from 3.2 to 3.4 m, in which sand from 10.3 to 11.4%, dust from 41.1 to 50.3%, clay from 40.3 to 50.2% and surface humidity from 48.1 to 49.5% may be broken connection between grains by wave with height under 0.2 m. However, for a wave with a height over 0.3 m, the shear stress of the wave is larger than the surface shear resistance of the clay layer. That proves that waves with a height of over 0.3 m may break the structure of the riverbank and lead to erosion.

In addition, the fluctuation of water level by seasons and tide caused riverbanks to be unstable and formed a bank scour mechanism. Materials of the surface layer of riverbanks are clay with depths from 3.2 to 3.4 m, in which sand grain occupied from 10.3 to 11.4%, dust grain from 41.1 to 50.3% and clay grain from 40.3 to 50.2%. This layer will be dried, hardened, and shrunk. These physical features will create cracks in the ground layer. Under the impacts of river flow, the ground will be unstable and eroded (Figs. 14.15 and 14.16).

14.5 Conclusions

1. Hau river is one of the main waterways and plays an important role in the socio-economic development of An Giang and as well as Mekong Delta. Besides natural impact factors causing riverbank unstable, human activities also affect riverbank erosion significantly.

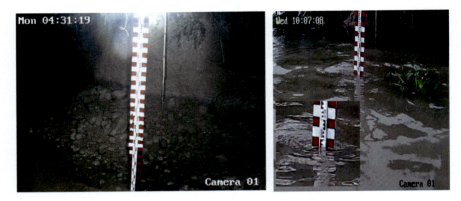

Fig. 14.15 Difference level of water by tide at My Hoa Hung isle

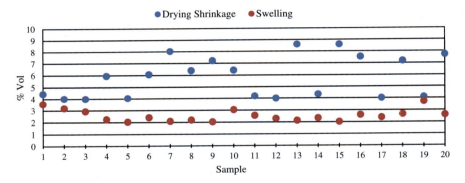

Fig. 14.16 Analysis result drying shrinkage and Swelling at My Hoa Hung and Binh Thanh isles

2. Recorded data from CCTV and AHW-10 USB wave gauge show that high vessel-generated waves increased high pressure on riverbanks, broken texture and eroded riverbanks.
3. Count and calculated the number of vessels through on the islet and how height waves are created by every single one.
4. Calculated the T_c is the critical shear stress through on Jet test in situ with remodeled samples have similar grain size distribution in Hau riverbank.
5. Through on Jet test, the Value of moisture content inverse ratio with τ_c. The τ_c value 12.81 will decrease suddenly to 2.80 N/m^2 when the moisture content increase from 45 to 47.8%, and got a value of 0.02 N/m^2 at 55.1% moisture content.
6. The majority of samples collected at two islets have a moisture content at the surface of more than 45% means the riverbank is easily destroyed by the wave. When the moisture is 47.8% with τ_c 2.80 is lower than bed shear stress (3.05 N/m^2) by a wave at 0.3 m height.

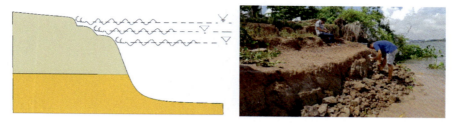

Fig. 14.17 Erosion mechanism of riverbanks in an Giang province, Vietnam

7. It is shown that under the impacts of natural and boat waves, materials of riverbanks are broken, abraded and transported to another place. This is a bank scour mechanism of the Hau riverbank and always happen strongly on the surface of riverbanks in high water period or tidal areas. The erosion mechanism of the Hau river is illustrated in Fig. 14.17.

References

Althage J (2010) Ship-induced waves and sediment transport in Göta river, Sweden. TVVR 10/5021

Hanson GJ (2001) Field and laboratory jet testing method for determining cohesive material erodibility. In: Proceedings of the seventh Federal interagency sedimentation conference

Hanson GJ, Cook KR (1997) Development of excess shear stress parameters for circular jet testing. ASAE Paper, 972227

Hanson GJ, Cook KR (2004) Apparatus, test procedures, and analytical methods to measure soil erodibility in situ. Appl Eng Agric 20(4):455

Le Dien T, Hoang NB, Vinh BT, Ngo DV (2012) The erosion state of Hau riverbank in an Giang province, south of Vietnam and recommended solutions for riverbanks protection. In: Proceedings of the international workshop, Hue Geo engineering 2012, pp 101–108

Roulund A, Sutherland J, Todd D, Sterner J (2016) Parametric equations for Shields parameter and wave orbital velocity in combined current and irregular waves. In: Scour and Erosion: proceedings of the 8th international conference on Scour and Erosion. CRC Press, Oxford, UK, p 313

Soulsby R (1988) Dynamics of marine sands, Thomas Telford

Vinh BT (2009) Estimation of erosion resistance of cohesive bank in river and around river mouth. Doctoral dissertation, Osaka University, Japan

Voulgaris G, Wallbridge S, Tomlinson B, Collins MJCe (1995) Laboratory investigations into wave period effects on sand bed erodibility, under the combined action of waves and currents. 26(3-4):117–134

Chapter 15
Sustainability and Disaster Mitigation Through Cascaded Recycling of Waste Tires—Climate Change Adaptation from Geotechnical Perspectives

Hemanta Hazarika, Yutao Hu, Chunrui Hao, Gopal Santana Phani Madabhushi, Stuart Kenneth Haigh, and Yusaku Isobe

15.1 Introduction

Concerns about climate change-related disasters linked to global warming have recently grown, and interventions and efforts to reduce CO_2 emissions have been introduced through the cooperation of industry, government, and academia. One of these methods is to recycle waste from human or manufacturing activities as geomaterials. Cascaded Recycling of such items, rather than simple recycling, is now promoted because it results in a significant reduction in CO_2 emissions to the atmosphere. In the case of waste tire recycling, Fig. 15.1 illustrates the principle of cascaded recycling addressed by Hazarika et al. (2018). In Japan, the total end-of-life tires produced by replacing tires and the quantity generated by scrapped vehicles was 96 million tires and over 1 million tons by weight in 2019. The report from JATMA (Japan Automobile Tyre Manufacturers Association 2020) shows that at least 61% of the scrap tires were used for energy recovery (thermal recycling), as shown in Fig. 15.2. The rest were reused either through returning the same product (material recycling) or through returning different products (cascaded recycling).

Waste tire statistics from around the world show a similar image. Every year, approximately 1 billion waste tires are produced in various parts of the world. The majority of these waste tires are either burned for energy or dumped on the ground. Thermal recycling, on the other hand, is detrimental to the atmosphere because it

---

H. Hazarika (✉) · Y. Hu · C. Hao
Kyushu University, Fukuoka 819-0373, Japan
e-mail: hazarika@civil.kyushu-u.ac.jp

G. S. P. Madabhushi · S. K. Haigh
University of Cambridge, Cambridge CB2 1PZ, UK

Y. Isobe
IMAGEi Consultant Corporation, Tokyo 102-0083, Japan

© The Author(s), under exclusive license to Springer Nature Singapore Pte Ltd. 2023
H. Hazarika et al. (eds.), *Sustainable Geo-Technologies for Climate Change Adaptation*, Springer Transactions in Civil and Environmental Engineering,
https://doi.org/10.1007/978-981-19-4074-3_15

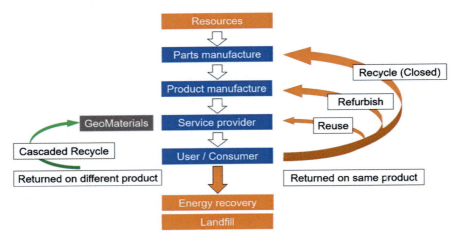

Fig. 15.1 Concept of cascaded recycling of waste tires (Hazarika et al. 2018)

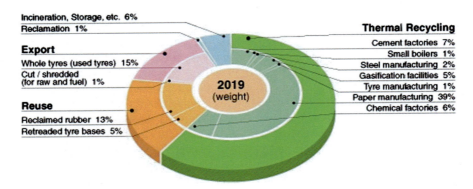

Fig. 15.2 Status of recycling of used tires in Japan as of 2019 (JATMA 2020)

emits more CO_2 than other recycling methods like material recycling and cascaded recycling. In order to promote waste tire recycling and reduce CO_2 emissions, the use of such industrial by-products in cascaded form (returning on different products) has gained popularity in recent years. As a result, it is important to change the recycling share from thermal to cascaded recycling. Scrap tire-derived materials (STDMs) are so called as they are recycled in a cascaded form. In addition to its low-carbon-release characteristics when used as geomaterials, other advantageous material characteristics of STDM include lightweight, excellent vibration absorption capability, and high permeability. Furthermore, unlike other granular geomaterials, these materials are non-dilatant in nature (Hazarika 2013). They can replace other traditional materials (such as gravel) in applications such as drainage, leachate removal in landfills, soil reinforcement, and so on because of these characteristics.

Many researchers in Japan have conducted a wide range of research on waste tire utilization. A few examples of the studies include: the use of whole tires or the use of tires in conjunction with other granular materials (Fukutake and Horiuchi 2006) or the use of tire chips and tire shreds (Hazarika et al. 2010, 2012a, b; Niiya et al. 2012; Karmokar et al. 2006; Kikuchi et al. 2008). In addition, tire chips mixed with cement treated clay have been used as sealing material at waste disposal sites in Tokyo Bay (Mitarai et al. 2006). The application of tire chips as a material to prevent liquefaction in soil has also been conducted by Yasuhara et al. (2010) and Uchimura et al. (2008). Studies on the physical and mechanical behavior of soil-tire-derived materials have shown that the shear strength behavior, dilatancy, and volumetric response of the mixtures are highly influenced by their tire content (Chiaro et al. 2019a, b; Pasha et al. 2019). Gravel-tire chips mixture (GTCM), as an alternative geomaterial, has been introduced by Hazarika et al. (2016). With a suitable gravel fraction by volume in the mixture, it is expected to be more practical and provide enough shear strength with higher permeability in comparison to that of the sand-tire chips mixture (Hazarika et al. 2020). Some studies have shown that GTCM is useful in reducing settlement and preventing the leakage of harmful leachate from marine landfill sites while using it as a drainage material in embankment construction and landfill leachate collection system (Hao et al. 2019, 2021a, b).

Large-scale earthquake-induced hazards or climate change-related hazards, on the other hand, have been seen around the world in recent years. One of such natural disasters, liquefaction, has become much too common in recent years, especially in Japan. According to the results compiled by the Ministry of Land, Infrastructure, Transport and Tourism, in the 2011 Tohoku Earthquake, about 27,000 houses were damaged due to liquefaction, about half of which were located in the Tokyo Bay area. Liquefaction-induced damage was also observed over a wide area following the 2016 Kumamoto Earthquake. Several areas in the southern part of Kumamoto City experienced liquefaction causing damage to residential houses due to differential settlement (Hazarika et al. 2017). The extreme damage to buildings caused by liquefaction during previous earthquakes has highlighted the value of taking preventive steps to protect buildings and facilities by reducing ground settling and lateral spreading. In addition, appropriate and cost-effective disaster mitigation efforts are desperately needed in most developing countries, where infrastructure growth is still in its infancy. In most developing countries, the main problem is striking a balance between the cost and the increased environmental impact of any infrastructure project.

One such low-cost technique was developed, which utilizes a layer of tire chips as the horizontal inclusion under the foundation of residential housing (Hazarika et al. 2009). Horizontal reinforcing inclusion refers to a layer of tire chips which is placed horizontally. GTCM was then used to replace the pure tire chips, as shown in Fig. 15.3. Since GTCM can provide sufficient bearing capacity to the foundation that otherwise has to rest on a highly compressible layer of tire chips, the improved technique is more practical (Hazarika et al. 2020). However, these mentioned techniques, as well as most of the other liquefaction countermeasures, are developed for new constructions. Very few techniques have been developed for existing buildings to mitigate liquefaction.

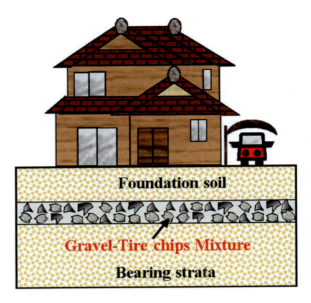

Fig. 15.3 Prevention of liquefaction-induced damage to building using horizontal inclusion (Hazarika et al. 2020)

The vertical drain method is one of the effective liquefaction countermeasures, that has been widely used in practice. The principal objective of using vertical drains is to relieve the excess pore water pressure generated during the shaking before they reach high values that can finally cause damage and loss to infrastructures (Brennan and Madabhushi 2006). More recently, Garcia-Torres and Madabhushi (2019) have investigated the performance of the drains underneath structures. Considering the performance of GTCM in horizontal reinforcing inclusion mentioned before, it could be possible to utilize this material in the vertical drain method as well. However, the permeability of GTCM should be analysed in advance.

In this paper, a series of large-scale triaxial compression and permeability tests were conducted. Under different levels of compression, the drainage performances of GTCM were examined. Using the results as a benchmark, a series of 1 g shaking table tests were performed to evaluate the performance of GTCM drains, as a new liquefaction mitigating technique. The aim is to develop a low-cost and sustainable solution for existing constructions to prevent liquefaction and its damage.

15.2 Tire-Derived Geomaterials as a Drainage Material

To utilize tire-derived geomaterials in the field construction, a comprehensive experimental study on permeability was necessary to perform ahead. This research examined the drainage performances of tire-derived geomaterials under different levels of compression which were largely unexplained until now. Herein, it should stress that after compression, drainage potentiality may be altered with disparate porosity.

Therefore, evaluating the drainage properties of tire-derived geomaterials essentially requires consideration of the compression behavior. In this research, a series of experimental studies on permeability and compressibility using tire-derived geomaterials (tire chips and gravel-tire chips mixture) was conducted as follows.

To evaluate the mechanical properties and permeability of large particle materials, new large-scale experimental apparatus was developed by Hazarika laboratory, Kyushu University. As shown in Fig. 15.4, the newly developed testing system for the triaxial compression and permeability test mainly includes an air cylinder (Load range 20 kN), pressure gage (Load range 50 kN), hydraulic jack (Load range 50 kN), constant hydrostatic head appliance, and water collection cylinder. Specimens are prepared in a large size mold with an inner diameter of 153 mm and a height of 300 mm. The thickness of the rubber membrane used in the test is 1.5 mm.

This large-scale triaxial compression and permeability testing apparatus is an extension of a traditional triaxial testing machine in that large particle material can be tested for compressibility and permeability simultaneously under different lateral pressure and water pressure conditions.

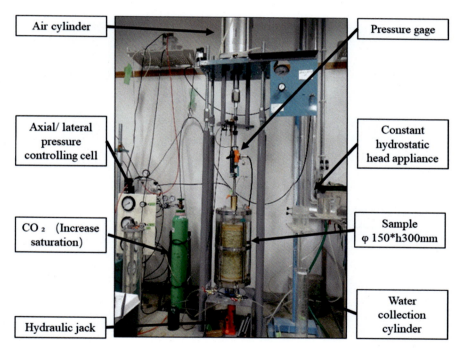

Fig. 15.4 Large-scale triaxial compression and permeability testing apparatus (Hao et al. 2021a, b)

15.2.1 Large-Scale Triaxial Compression and Permeability Tests

Tire chips and gravel samples used in this study are shown in Fig. 15.5. Compared to tire shreds, almost all steel and textile cords were derived from tire chips. The grain size of tire chips is in the range of 9.5–19 mm and for gravel, it is 9.5–15 mm. The particle density of gravel is 2.64 g/cm^3, tire chips have a particle density of 1.14 g/cm^3, less than tire shreds. However, based on the high compressibility of tire chips, gravel can be considered as an acceptable material for the mixture in this research.

Based on previous studies about GTCM by Pasha et al. (2019) and Chu et al. (2018), tire chips contents of 0 (pure gravel), 50% (gravel-tire chips mixture, GTCM), and 100% (pure tire chips) by weight and relative density of 70% are selected in this research. After completion of the isotropic consolidation process, the axial load is increased by a multiple lateral pressure. As shown in Fig. 15.6, to evaluate the drainage behavior simultaneously, one hour for each increase in pressure should be

Tire chip (9.5-19mm) Gravel (9.5-15mm)

Fig. 15.5 Gravel, and tire chips used in the experimental study

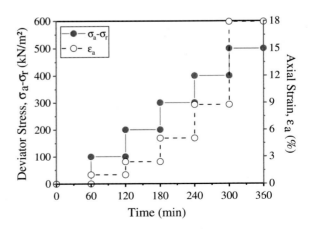

Fig. 15.6 Stress loading and axial strain (Gravel, $\sigma_r = 100$ kPa)

paused. As a traditional triaxial test, axial compression will be stopped when the axial strain of 15% has been reached. The permeability test is performed only when considerable strain producing. It is worth mentioning that the hydraulic gradient is ranged from 0.2 to 1, because when the sample changes significantly after axial load, the corresponding head difference will be small, which is difficult to adjust by this apparatus.

15.2.2 Test Results

The influence of MPTC (mass proportion of tire chips) and lateral pressure on the deviatoric stress-axial strain relationship is shown in Fig. 15.7. Unlike gravel particles, tire chip particles are easily deformed. For MPTC is 50–100%, based on the low stiffness of tire chip particles, a significant reduction in low shear strength is shown in pure tire chips and GTCM (gravel-tire chips mixture, MPTC = 50%) specimens. It can be obviously seen that the completely linear behavior of the deviatoric stress-axial strain relationship in the axial compression process. From Fig. 15.16, the MPTC = 100%, $\sigma_r = 150$ kN/m^2 and MPTC = 50%, $\sigma_r = 100$ kN/m^2, shear stress behaviors are almost same. For MPTC $\geq$ 50%, a stress path similar to gravel specimens has not been found in this experiment.

The permeability coefficients of GTCM under different MPTC and pressure conditions are illustrated in Figs. 15.8 and 15.9. As shown in Fig. 15.8, with the increase of

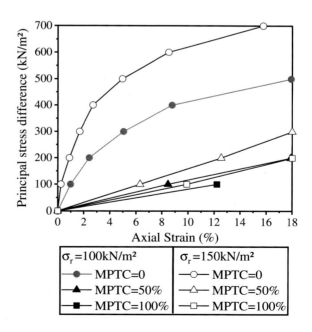

Fig. 15.7 Influence of MPTC (Mass proportion of tire chips) and lateral pressure on deviatoric stress-axial strain relationship

Fig. 15.8 Influence of MPTC on permeability coefficient (Isotropic pressure state)

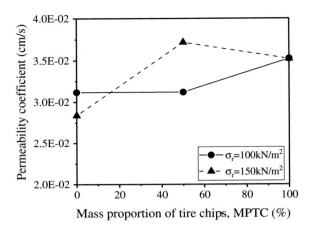

Fig. 15.9 Influence of MPTC and lateral pressure on permeability coefficient-axial strain relationship

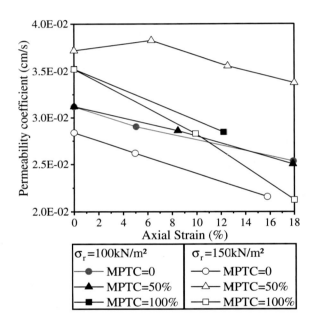

the MPTC, the permeability coefficient showed an increasing trend under isotropic pressure. In other words, the permeability of GTCM is promoted by tire chips content. As shown in Fig. 15.9, GTCM has an excellent permeability coefficient on the order of 0.02–0.04 cm/s under different axial pressure. Herein, the height of the sample (MPTC = 100%, $\sigma_r = 100$ kN/m^2) changed greatly after axial pressure, the maximum axial strain of 12% is the axial length limitation of this equipment. In this experimental study, GTCM sample (MPTC = 50%, $\sigma_r = 150$ kN/m^2) exhibits the maximum permeability coefficient. It can be explained that the stiffness and permeability are increased through the mixing of gravel and tire chips. In summary, the permeability of tire chips and GTCM is similar to that of conventional coarse gravel.

These results provided the initial findings regarding the permeability and compressibility of tire-derived geomaterials like tire shreds, tire chips, and GTCM (gravel-tire chips mixture). To evaluate the best combination of GTCM as a drainage material, the mass proportion of tire chips is also an important design consideration in this research.

15.3 Performance of Gravel-Tire Chips Mixture Drains

Figure 15.10 describes the concept of this new technique, by using GTCM drains, to reduce liquefaction potential for existing buildings. A series of 1 g shaking table tests were conducted at the geo-disaster laboratory of Kyushu University to evaluate the effectiveness of GTCM drains in dissipating excess pore water pressure and reduce liquefaction potential during an earthquake.

15.3.1 Test Cases and Test Producers

Models were constructed in a transparent Plexiglas container with dimensions of 1800 mm × 400 mm × 850 mm, as shown in Fig. 15.11. Toyoura sand was used as foundation soil in these tests. A dense layer of such sand ($D_r = 90\%$) representing non-liquefiable ground was constructed using both dry deposition and tamping techniques. The depth of the dense layer was 200 mm. The upper liquefiable layer ($D_r = 50\%$) with a depth of 300 mm was constructed only using the dry deposition technique. The properties of the material are shown in Table 15.1. Since heavy building could settle more due to its weight rather than the influence of liquefaction (Hu et al.

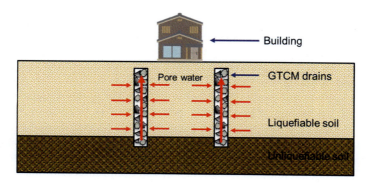

Fig. 15.10 GTCM drains technique

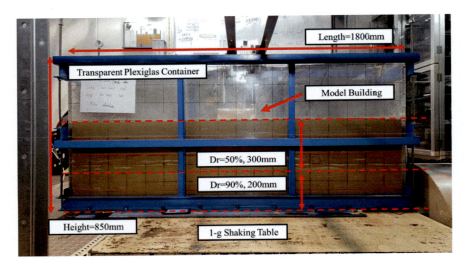

Fig. 15.11 1 g shaking table test model

Table 15.1 Properties of Toyoura sand

Property	Value
G_s	2.640
Dry density (g/cm^3)	1.506
e_{max}	0.976
e_{min}	0.611
D_{50} (mm)	0.160

2021), a shallow foundation with a bearing pressure of 3 kPa represented by rectangular blocks of brass material was modeled, with a cross-sectional area of 230 mm × 100 mm in model scale. The saturation process could be performed by percolating water gradually and uniformly through 3 water inlets from the bottom of the container.

Soil-structure-fluid interaction can be simulated using the scaling law proposed by Iai (1989). Since this research involves liquefaction-induced damage to the structure, it is the most suitable similitude relationship, and, therefore this similitude law was used in this research. Throughout these tests, a geometrical scaling factor of 1:32 was set based on this law, as shown in Table 15.2.

Depending on whether GTCM drains were installed in the foundation or not, two cases of 1 g shaking table tests were performed. Case 1 was set as the default condition, with no drains installed. As compassion, in Case 2, 4 parallel arrays of prefabricated GTCM drains with a diameter of 50 mm and height of 300 mm were installed vertically around the four sides of a residential building from the surface level up to the bottom of the loose sandy layer and extended into the hard layer.

15 Sustainability and Disaster Mitigation Through Cascaded Recycling ...

Table 15.2 The scaling law used in the tests

Parameter	Scaling law (prototype/model)
Length	$n = 32$
Density	1
Stress and pressure	$n = 32$
Time	$n^{0.75} = 13.5$
Frequency	$n^{-0.75} = 0.074$
Displacement	$n^{1.5} = 181$
Acceleration	1

A sinusoidal acceleration of 200 Gal with the frequency of 4 Hz and duration of 10 s was applied to the model in both two cases. Cross and top sectional views of the tests performed are shown in Figs. 15.12 and 15.13. Different types of transducers were employed to measure acceleration, pore water pressure and displacement at different positions, as shown in this figure. The pore pressure transducers (PPTs) were fixed in place to monitor the pore water pressure in the exact locations. The laser micro displacement transducers (LMDTs) were set at the middle of short sides on the top of the model buildings.

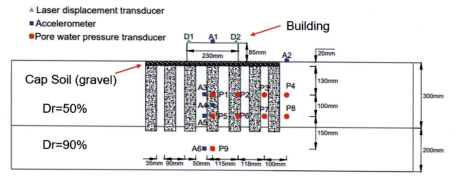

Fig. 15.12 Cross-section view of the model

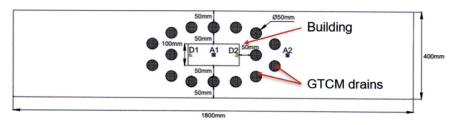

Fig. 15.13 Top-section view of the model

15.3.2 Test Results

Figures 15.14 and 15.15 show the time history of excess pore water pressure ratio defined as $R_u = \Delta u/\sigma'$ recorded by PPTs. In the without drains foundation (Case 1) as shown in Fig. 15.14, the excess pore water pressure ratio recorded exceeds 1.0 indicating liquefaction in the foundation soils, except the one directly beneath the building during the earthquake. The value of R_u initiates fast increasing as soon as the earthquake starts and keeps in high until the end of shaking. However, due to the compaction from the building, the increase of R_u beneath the building is much slower. While in the drains foundation (Case 2), no increase in excess pore water pressure was observed from Fig. 15.15. The results imply that GTCM drains helped in dissipating excess pore water pressure during the earthquake and therefore show the ability in preventing liquefaction.

Figure 15.16a shows the time history of acceleration for the building in Case 1 under the input excitation with an amplitude of 200 Gal. As can be seen from the figure, ground motion is amplified through the upward propagation within the soil from the bottom to the ground surface of the liquefiable soil (A5 to A2 shown in Fig. 15.12). The acceleration time histories also showed spikes in the liquefiable layer

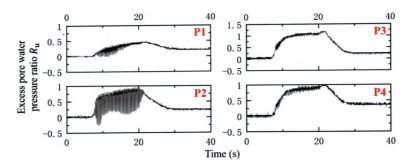

Fig. 15.14 Excess pore water pressure ratio in Case 1 (without drains)

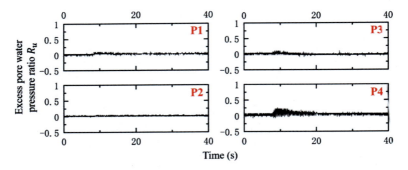

Fig. 15.15 Excess pore water pressure ratio in Case 2 (with drains)

15 Sustainability and Disaster Mitigation Through Cascaded Recycling ...

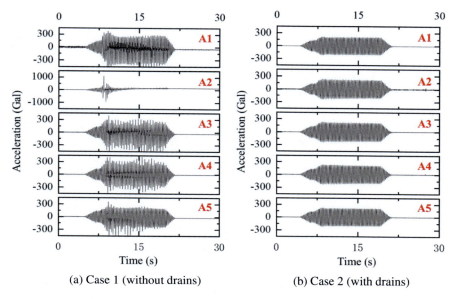

Fig. 15.16 Time history of acceleration

(A2, A3 and A4 shown in Fig. 15.12). Larger acceleration spikes were observed as the density of soil increased due to subsequent shaking, which may be attributed to soil softening and re-stiffening induced by a high level of excess pore water pressure. The high level of excess pore water pressure induced softening in the soil in the light building case, resulting in a disassociation between input motion and acceleration of soil.

Figure 15.16b shows the time history of acceleration for the building in Case 2 under the input excitation with an amplitude of 200 Gal. No amplification of the ground motion was observed through the whole foundation from the bottom to the surface.

The settlement time histories at the same side of the building, recorded by LMDT D2 (shown in Fig. 15.13), are presented in Fig. 15.17. The maximum settlement of the building on unimproved soil is around 32 mm. While on improved soil, the building only settled less than 0.88 mm. The difference of 36 times implies the effectiveness of GTCM drains in preventing liquefaction-induced settlement to the existing building.

Figure 15.18 shows the displacement of the model building after the earthquake in Case 1. This figure showed significant settlement. The building sank to the original ground level and nearly disappeared after the liquefaction event. The rotation of the building was also obvious as shown in the figure. In opposite, with GTCM drains installed in Case 2, the settlement of the model building was too small to be observed directly, as shown in Fig. 15.19.

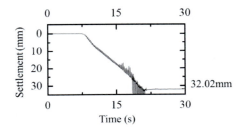

(a) Settlement at D2 in Case 1 (without drains)

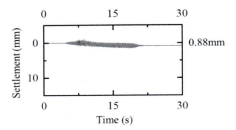

(b) Settlement at D2 in Case 2 (with drains)

Fig. 15.17 Time history of settlement at D2

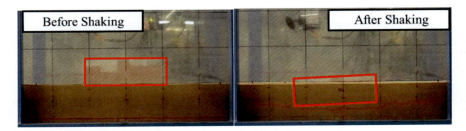

Fig. 15.18 The displacement of model building in Case 1

Fig. 15.19 The displacement of model building in Case 2

15.4 Concluding Remarks

The following are some of the main conclusions derived from this research.

1. The GTCM has excellent permeability of 0.02–0.04 cm/s under different triaxial strains, and it seems that the permeability of GTCM is almost the same as gravel drainage material.
2. The stiffness and permeability are increased through the mixing of gravel and tire chips (GTCM). However, the optimum combination of GTCM is an issue that still needs to be considered in long-term drainage performance.
3. With the GTCM drains installed, the excess pore water can dissipate through the drains from the foundation to the surface of the soil fast. As a result, the potential of liquefaction around the existing buildings would be reduced.
4. The high level of excess pore water pressure induced softening in the soil in the light building case, resulting in a disassociation between input motion and acceleration of soil.
5. Liquefaction-induced settlement of the building can be controlled to a low level during the earthquake since the liquefaction would be prevented through GTCM drains.

Acknowledgements The authors would like to acknowledge the financial support provided by Kyushu University under Progress 100 project. Special thanks go to Mr. Yuichi Yahiro, technical assistant of the Geo-disaster Laboratory of Kyushu University, for his help and support while conducting experiments.

References

Brennan AJ, Madabhushi SPG (2006) Liquefaction remediation by vertical drains with varying penetration depths. Soil Dyn Earthq Eng 26(5):469–475

Chiaro G, Palermo A, Granello G, Banasiak LJ (2019a) Direct shear behaviour of gravel-granulated tyre rubber mixtures. In: Proceedings of the 13th ANZ geomechanics conference, Perth Australia, pp 6

Chiaro G, Palermo A, Granello G, Tasalloti A, Stratford C, Banasiak LJ (2019b) Eco-rubber seismic-isolation foundation systems: a cost-effective way to build resilience. In: Proceedings of the 11 pacific conference on earthquake engineering, Auckland, NZ

Chu C, Hazarika H, Isobe Y (2018) Application of the waste tire cascade recycling to the seafloor protection at seashore landfill for final disposal. In: The 53rd Japan National Conference on Geotechnical Engineering, Takamatsu, Japan, CD-ROM (In Japanese)

Fukutake K, Horiuchi S (2006) Forming method of geostructure using recycled tires and granular materials. In: Proceedings of the 41st Japan National Conference on Geotechnical Engineering, pp 653–654 (in Japanese)

Garcia-Torres S, Madabhushi SPG (2019) Performance of vertical drains in liquefaction mitigation under structures. Bull Earthq Eng 17(11):5849–5866

Hao CR, Hazarika H, Isobe Y (2019) Applicability of tire derived geomaterials in marine landfill sites. In: Proceedings of the technical forum on mitigation of geo-disasters in Asia, Kumamoto, Japan, pp 120–123

Hao CR, Hazarika H, Isobe Y (2021a) Evaluation of gravel-tire chips mixtures for their use in marine landfill leachate collection systems. In: Proceedings of the 20th international conference on soil mechanics and geotechnical engineering, international society for soil mechanics and geotechnical engineering, Sydney

Hao CR, Hazarika H, Isobe Y (2021b) Performance assessment of recycled tire materials in marine landfill application. In: Hazarika H, Madabhushi GSP, Yasuhara K, Bergado DT (eds) Advances in sustainable construction and resource management, vol 144. Lecture Notes in Civil Engineering. Springer, Singapore, pp 117–126

Hazarika H (2013) Paradigm shift in earthquake induced geohazards mitigation—Emergence of nondilatant geomaterials. In: Keynote lecture for the annual conference of Indian geotechnical society, Roorkee, India, CD-ROM

Hazarika H, Abdullah A (2016) Improvement effects of two and three dimensional geosynthetics used in liquefaction countermeasures. Jpn Geotech Soc Special Publ 2(68):2336–2341

Hazarika H, Igarashi N, Yamagami T (2009) Evaluation of ground improvement effect of tire recycle materials using shaking table test. In: Proceedings of the 64th annual conference of Japan Society of Civil Engineers, pp 931–932 (In Japanese)

Hazarika H, Kokusho T, Kayen RE, Dashti S, Fukuoka H, Ishizawa T, Kochi Y, Matsumoto D, Furuichi H, Hirose T, Fujishiro T, Okamoto K, Tajiri M, Fukuda M (2017) Geotechnical damage due to the 2016 Kumamoto earthquake and future challenges. Lowland Technol Int 19(3):189–204

Hazarika H, Otani J, Kikuchi Y (2012a) Evaluation of tyre products as ground improving geomaterials. In: Ground improvement, Institution of Civil Engineers, vol 165, no GI1, UK, pp 1–16

Hazarika H, Pasha SMK, Ishibashi I, Yoshimoto N, Kinoshita T, Endo S, Karmokar AK, Hitosugi T (2020) Tire chip reinforced foundation as liquefaction countermeasure for residential buildings. Soils Found 60(2):315–226

Hazarika H, Yasuhara K, Kikuchi Y, Karmokar AK, Mitarai Y (2010) Multifaceted potentials of tire-derived three dimesional geosynthetics in geotechnical applications and their evaluation. Geotext Geomembr 28(3):303–315

Hazarika H, Yasuhara K, Kikuchi Y, Kishida T, Mitarai Y, Sugano T (2012b) Novel earthquake resistant reinforcing technique (SAFETY) using recycled tire materials. Geotech Eng Mag Jpn Geotech Soc 60(9):30–31 (in Japanese)

Hazarika H, Yokota H, Endo S, Kinoshita T (2018) Cascaded recycle of waste tires—Some Novel approaches toward sustainable geo-construction and climate change adaptation. In: Krishna A, Dey A, Sreedeep S (eds) Geotechnics for natural and engineered sustainable technologies. Developments in geotechnical engineering. Springer, Singapore

Hu Y, Hazarika H, Pasha SMK, Haigh SK, Madabhushi GSP (2021) Effect of bearing pressure on liquefaction-induced settlement in layered soils. In: Hazarika H, Madabhushi GSP, Yasuhara K, Bergado DT (eds) Advances in sustainable construction and resource management. Lecture notes in civil engineering, vol 144. Springer, Singapore, pp 261–270

Iai S (1989) Similitude for shaking table tests on soil-structure-fluid model in 1g gravitational field. Soils Found 29(1):105–118

Japan Automobile Tyre Manufacturers Association (2020). Tyre industry of Japan. http://www.jatma.or.jp/media/pdf/tyre_industry_2020.pdf

Karmokar AK, Takeichi H, Kawaida M, Kato Y, Mogi H, Yasuhara K (2006) Study on thermal insulation behavior of scrap tire materials for their use in cold region civil engineering applications. In: Proceedings of the 60th Japan society of Civil Engineers annual meeting, Tokyo, Japan, pp 851–852

Kikuchi Y, Sato T, Nagatome T, Mitarai T, Morikawa Y (2008) Change of failure mechanism of cement treated clay by adding tire chips. In: Proceedings of the 4th Asian Regional conference on geosynthetics, Shanghai, pp 374–379

Mitarai Y, Yasuhara K, Kikuchi Y, Karmokar AK (2006) Application of cement treated clay added with tire chips to the sealing materials of coastal waste disposal site. In: Proceedings of the 6th international congress on environmental geotechnology, vol 1, Cardiff, UK, pp 757–764

Niiya F, Hazarika H, Yasufuku N, Ishikura R (2012) Cyclic frictional behavior of two and three dimensional geosynthetics used in liquefaction countermeasure. In: Proceedings of the 5th Taiwan-Japan joint workshop on large earthquakes and heavy rainfall, Tainan, Taiwan, CD-ROM

Pasha SMK, Hazarika H, Yoshimoto N (2019) Physical and mechanical properties of gravel-tire chips mixture (GTCM). Geosynth Int 26(1):92–110

Uchimura T, Chi NA, Nirmalan S, Sato T, Meidani M, Towhata I (2008) Shaking table tests on effect of tire chips and sand mixture in increasing liquefaction resistance and mitigating uplift of pipe. In: Hazarika, Yasuhara (eds) Scrap tire derived geomaterials—Opportunities and challenges. Taylor & Francis Group, London, pp 179–186

Yasuhara K, Komine H, Murakami S, Miyota S, Hazarika H (2010) Mitigation of liquefaction using tire chips as a gravel drain. In: Proceedings of the 6th International Congress on Environmental Geotechnics, India, CD-ROM

Printed in the United States
by Baker & Taylor Publisher Services